"EVALUATING THE MEASUREMENT PROCESS"

SECOND EDITION

DONALD J. WHEELER
RICHARD W. LYDAY

SPC PRESS, INC.

KNOXVILLE, TENNESSEE

Copies may be purchased from
SPC Press, Inc.
5908 Toole Drive, Suite C
Knoxville, Tennessee 37919
(615) 584-5005

ISBN 0-945320-06-X   3 4 5 6 7 8 9 10

CONTENTS

PART TWO: EMP CASE HISTORIES

The Four Foundations
of
Shewhart's Control Charts

I.   Shewhart's Control Charts will always use control limits which are set at a distance of three standard deviations on either side of the central line.

II. Every standard deviation used to compute these limits must be estimated by using the average variation within the subgroups.[*] No other estimates of the standard deviations are acceptable.

III. The data shall be obtained by a rational sampling scheme and shall be organized into subgroups in a rational manner. The organization of the data into subgroups must respect both the context of the data and the questions to be addressed by the charts.

IV. The organization can and will utilize the knowledge gained from the charts.   An organization must have the ability to respond to knowledge by changing the organization's actions as appropriate before Shewhart's control charts will be as effective as they can be.

---

[*] When the rational subgroup size is n = 1 one can no longer compute an ordinary within-subgroup estimate of the standard deviation of X. However, in the interest of robustness, one still **cannot** justify using the overall standard deviation for the set of k values. Instead one will need to use the two-point moving range. This statistic will provide the robust estimate required for the control charts to be fully effective. This technique is illustrated on page 16.

Preface

The popularity of the first edition of this book provided us with many opportunities to discuss the evaluation of measurement processes with others. Based on these discussions we came to feel the need to explain more fully the ways in which control chart techniques may be applied to such evaluations. This second edition is the result.

This book assumes that the reader has some familiarity with control charts. The methods and procedures outlined herein are merely applications of the Foundations of Shewhart's Control Charts. In particular, this book is an extended explanation of how to use the Third Foundation in different situations in order to address specific questions of interest. The techniques in this book have developed from the premise that "a measurement process is simply a production process which produces numbers."

Two new procedures are introduced in this edition. They are the Chart for Main Effects, and the Chart for Mean Ranges. While both of these charts are logical extensions of the control chart technique, they have not been widely described or used. For this reason, worksheets are provided for both of these procedures in the Appendix.

Throughout this book we have attempted to keep the discussion focused upon real world issues and real world solutions. This is not intended to be a theoretical book, therefore proofs for different procedures are not provided. Rather, this book is intended to be a users' manual for industrial personnel who have to make decisions based upon imperfect data. To the extent that we may succeed in this, we are indebted to the many different users of the first edition who have pooled their experiences with ours. Without the practical and applied work in the industrial context engendered by the first edition, this second edition would not have been possible.

In addition to those mentioned above, we are espeically indebted to Dr. James Maynard for his many thoughtful and helpful suggestions.

Donald J. Wheeler
Richard W. Lyday
August, 1989

"An element of chance enters into every measurement;
hence every set of measurements
is inherently a sample of certain more or less unknown conditions.
Even in those few instances where we believe
that the objective reality being measured is a constant,
the measurements of this constant are influenced by chance or unknown causes."

W.A. Shewhart

No two things are alike, but even if they were,
we would still get different values when we measured them.

Introduction

Questions Regarding Measurements

The key to the effective use of any measurement is an understanding of the different sources of variation contained in that measurement. Since many of these sources of variation are connected with the measurement process itself, it will often be profitable to study that process. Simple and effective techniques for doing this are presented in this book. They are organized around the following questions about measurements.

1. Are the measurements being made with measurement units which are small enough to properly reflect the variation present? When the measurement units are too large, the measurements will display one type of inadequate discrimination. Regular control charts for a product or a process indicate when this problem occurs.

2. Are the measurements consistent over time? The only way to answer this question is to repeatedly measure things which are thought to be the same. An operational definition of consistent measurements is given in Chapter Two.

3. Are the measurements biased? In an absolute sense, this question is difficult to answer. When an answer is possible, control charts provide a simple means of obtaining the answer.

4. How much uncertainty should be attached to a measurement when interpreting it? Just what is the effective resolution of a measurement? The answer to this question requires a way to quantify measurement uncertainty. Methods for using control charts to find just such a measure are outlined in Chapters Three and Four.

5. How can a single reading, or the average of several readings, be used to characterize product relative to specification limits? A method of interpreting a measurement relative to specification limits is outlined in Chapter Three.

6. Can one use measurements to characterize product which has not been measured? Can one meaningfully "accept" or "reject" batches of product based upon one or more measurements? The implicit assumptions for such a course of action are discussed in Chapter Three.

7. When can one use measurements to make adjustments to a production process? The conditions and assumptions implicit in this common usage of measurements are discussed in Chapter Three.

8. Do the measurements display detectable differences from operator to operator, machine to machine, lab to lab, or day to day? Of course one must have data collected by several operators, on several machines, etc., and these data will have to be obtained in such a way to permit comparisons. Simple methods for answering these questions are presented in Chapter Four.

9. How can one compare different techniques of measuring a particular characteristic? Simple methods for comparing different measurement techniques are outlined in Chapters Four and Five.

10. How much information about the product is contained within the measurements? How much of the natural variation is due to measurement error? Can we detect process improvement if and when it happens? All of these questions are concerned with the ability of the measurement process to detect product variation within the Natural Process Limits. A simple control chart can provide a graphic answer, and the data from this control chart can be used to obtain a number which will quantify the relative usefulness of a particular measurement procedure for a given product.

These questions cover the most pressing issues concerning the measurement process. They are directly concerned with the utility of a measurement process in a specified context, i.e., for a specific product. Because the questions are focused in this manner, the answers will often clarify the issues, making significant improvements possible.

The issues of "linearity of accuracy" and "constancy of error" are missing from the list above. This omission was deliberate. These broader questions concern the behavior of the measurement process across a wide spectrum of applications. The techniques in this book focus upon the question of using a specific measurement for a specific product. This narrow focus was maintained because the questions of interest to most practitioners fall within the specific category. Very few care to know about the "goodness" of a measurement process over some range where it is not used. However, should it become necessary, the techniques given herein may be adapted to address such questions.

Finally, the student needs to know that this book contains two parts. Part One contains the descriptions of the different computations and outlines the logic behind these computations. Part Two consists of five extensive and detailed examples of EMP Studies.

Chapter One

Inadequate Measurement Units

One of the simplest measurement problems is the use of measurement units which are too large for the job. This problem is fairly widespread, and is easily detected by ordinary control charts for product measurements. No special studies are necessary; no standard parts or batches are needed. One simply needs to understand the tell-tale signs on a regular Average and Range Chart. It is the purpose of this chapter to explain the tell-tale signs of Inadequate Measurement Units, to outline the nature of the problem, why the problem exists, and what can be done about it when it occurs.

In order to do this one must answer two basic questions: What happens to the control chart when the measurement units used are too large? And, what is the effect on control charts of measurement "round-off"?

The easiest way to answer these two questions is by manipulating a data set to create measurement units which are too large. One may then compare the control charts before and after the manipulation of the data to discover the effect of excessively large measurement units.

The data in Table 1.1 are the measurements of a physical dimension on a plastic knob. The data are recorded in inches, but the smallest measurement unit is one/one-thousandth of an inch (0.001 in.).

Table 1.1: Rheostat Knob Data

Subgroup						X̄	R	Subgroup						X̄	R
1	.140	.143	.137	.134	.135	.1378	.009	15	.144	.142	.143	.135	.144	.1416	.009
2	.138	.143	.143	.145	.146	.1430	.008	16	.133	.132	.144	.145	.141	.1390	.013
3	.139	.133	.147	.148	.149	.1432	.016	17	.137	.137	.142	.143	.141	.1400	.006
4	.143	.141	.137	.138	.140	.1398	.006	18	.137	.142	.142	.145	.143	.1418	.008
5	.142	.142	.145	.135	.136	.1400	.010	19	.142	.142	.143	.140	.135	.1404	.008
6	.136	.144	.143	.136	.137	.1392	.008	20	.136	.142	.140	.139	.137	.1388	.006
7	.142	.147	.137	.142	.138	.1412	.010	21	.142	.144	.140	.138	.143	.1414	.006
8	.143	.137	.145	.137	.138	.1400	.008	22	.139	.146	.143	.140	.139	.1414	.007
9	.141	.142	.147	.140	.140	.1420	.007	23	.140	.145	.142	.139	.137	.1406	.008
10	.142	.137	.134	.140	.132	.1370	.010	24	.134	.147	.143	.141	.142	.1414	.013
11	.137	.147	.142	.137	.135	.1396	.012	25	.138	.145	.141	.137	.141	.1404	.008
12	.137	.146	.142	.142	.146	.1426	.009	26	.140	.145	.143	.144	.138	.1420	.007
13	.142	.142	.139	.141	.142	.1412	.003	27	.145	.145	.137	.138	.140	.1410	.008
14	.137	.145	.144	.137	.140	.1406	.008								

The Average and Range Charts for these data are shown in Figure 1.1. There are no indications of a lack of control on either chart.

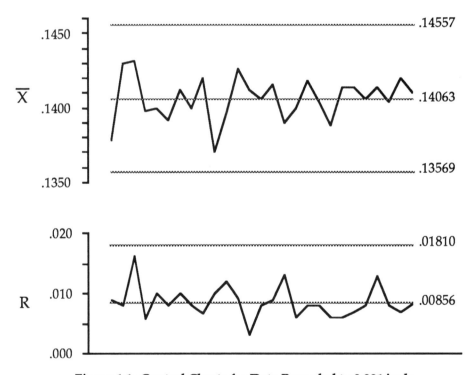

Figure 1.1: Control Charts for Data Recorded to 0.001 inch

The data in Table 1.2 were obtained from the data in Table 1.1 by rounding off each value to the nearest one/one-hundredth of an inch (0.01 in.). Values ending in 5 were rounded to the nearest even multiple of 0.01". Following this rounding off, the subgroup averages and ranges were re-computed, yielding the values shown.

Table 1.2: Rounded Data for Rheostat Knob

Subgroup						$\overline{X}$	R	Subgroup						$\overline{X}$	R
1	.14	.14	.14	.13	.14	.138	.01	15	.14	.14	.14	.14	.14	.140	.00
2	.14	.14	.14	.14	.15	.142	.01	16	.13	.13	.14	.14	.14	.136	.01
3	.14	.13	.15	.15	.15	.144	.02	17	.14	.14	.14	.14	.14	.140	.00
4	.14	.14	.14	.14	.14	.140	.00	18	.14	.14	.14	.14	.14	.140	.00
5	.14	.14	.14	.14	.14	.140	.00	19	.14	.14	.14	.14	.14	.140	.00
6	.14	.14	.14	.14	.14	.140	.00	20	.14	.14	.14	.14	.14	.140	.00
7	.14	.15	.14	.14	.14	.142	.01	21	.14	.14	.14	.14	.14	.140	.00
8	.14	.14	.14	.14	.14	.140	.00	22	.14	.15	.14	.14	.14	.142	.01
9	.14	.14	.15	.14	.14	.142	.01	23	.14	.14	.14	.14	.14	.140	.00
10	.14	.14	.13	.14	.13	.136	.01	24	.13	.15	.14	.14	.14	.140	.02
11	.14	.15	.14	.14	.14	.142	.01	25	.14	.14	.14	.14	.14	.140	.00
12	.14	.15	.14	.14	.15	.144	.01	26	.14	.14	.14	.14	.14	.140	.00
13	.14	.14	.14	.14	.14	.140	.00	27	.14	.14	.14	.14	.14	.140	.00
14	.14	.14	.14	.14	.14	.140	.00								

The Average and Range Charts for the data in Table 1.2 are shown in Figure 1.2. One now finds several indications of a lack of control, even though none were found in Figure 1.1.

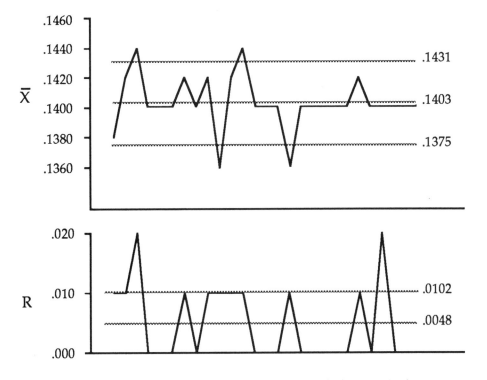

Figure 1.2: Control Charts for Data Recorded to 0.01 inch

The out-of-control points in Figure 1.2 have nothing to do with the underlying physical process. Instead, they were created by the excessive round-off present in Table 1.2.

Thus, excessive round-off in the measurements can make a control chart appear to be out-of-control even when the underlying process is in a state of statistical control.

Fortunately, when this problem exists, it is easy to identify. Ordinary control charts signal the presence of inadequate discrimination due to measurement units which are too large: they do this by showing too few possible values within the control limits on the range chart.

If there are only 1, 2, or 3 possible values for the range within the control limits, then the measurement units are too large for the purposes of the control chart. Moreover, if the subgroup size is three or larger ($n \geq 3$), and there are only 4 possible values for the range within the control limits, then the measurement units are too large for the purposes of the control chart.

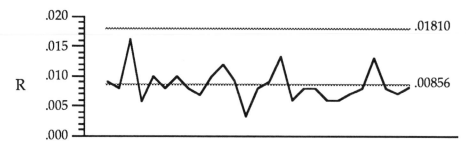

Figure 1.3: Range Chart for Data Recorded to 0.001 inch

In Figure 1.3, there are 19 possible values for the range within the control limits. This is evidence that there is no problem with Inadequate Measurement Units.

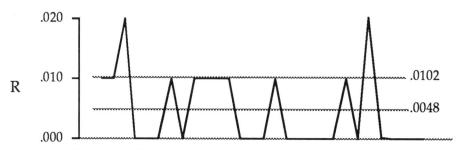

Figure 1.4: Range Chart for Data Recorded to 0.01 inch

In Figure 1.4, there are only two possible values for the range within the control limits. This is sufficient evidence that these charts suffer from Inadequate Measurement Units. In other words, the measurement units are too large for the purposes of the chart. When this happens, one cannot safely interpret the control chart as reflecting the behavior of the process. (Any out-of-control points may be due to inadequate discrimination.) This problem must be corrected before the control chart will be of any real use.

The problem seen on the control charts in Figures 1.2 and 1.4 is due to the inability of the measurements to properly detect and reflect the process variation. When the measurements are rounded off to the nearest 0.01 inch, most of the information about dispersion is lost in the round-off. As a result, while the original data have no zero ranges, the rounded data have many zero ranges. These zero ranges deflate the Average Range and tighten the control limits. At the same time, the greater discreteness for both the averages and the ranges will spread out the running records. Eventually it becomes inevitable that some points will fall outside the limits, even though the process itself is **not** out-of-control.

Since this problem arises out of the inability to detect variation within the subgroups, the solution consists of increasing the ability to detect that variation. This can be done in one of two ways. **Either use smaller Measurement Units, or increase the variation within the subgroups to a detectable level.** The latter approach will usually be accomplished by collecting values for a subgroup over a longer time span. By increasing the time span, one often increases the variation enough to make it detectable with the original Measurement Units.

Detection Rules for Inadequate Measurement Units

The justification of the rules for detecting Inadequate Measurement Units is based upon the relationship between the smallest actual unit of measurement and the process standard deviation.

To understand this relationship, consider a subgroup of size n = 2. The Average Range for such subgroups will be approximately 1.128 SD(X) [1.128 times the Standard Deviation of the Distribution of X]. Since the range is just the distance between the two measurements in a subgroup, and since any collection of measurements from a stable process could be arbitrarily arranged into subgroups of size n = 2, we can say that the average distance between any two randomly selected measurements is about 1.1 SD(X). (1.128 is the value for d_2 for subgroups of size 2.)

At the same time, consider two measurements which are rounded off to the same value: two measurements will be rounded to the same value when they both differ from that value by less than one-half measurement unit. For example, in measuring to the nearest millimeter, the measurements of 14.51 mm and 15.49 mm will both be rounded off to the value of 15 mm. Thus, measurements might be rounded to the same value when they differ by less than one measurement unit.

The observations in the two preceding paragraphs can be combined to discover the root of the problem created by Inadequate Measurement Units. When the Process Standard Deviation, SD(X), is smaller than the Measurement Unit, the measurements will begin to be rounded to the same values. This rounding will begin to contaminate the estimate of the Process Standard Deviation by deflating the Average Range. Moreover, the smaller the Process Standard Deviation is relative to the Measurement Unit, the greater the contamination will be.

Therefore, the problem of Inadequate Measurement Units (inadequate discrimination due to a measurement unit which is too large) begins to affect the control chart when the Measurement Unit exceeds the Process Standard Deviation.

The control charts are on the borderline of Inadequate Measurement Units when the Process Standard Deviation is equal to the Measurement Unit. The rule for detecting this problem is based upon the borderline condition. The origin of this rule is outlined below.

Consider the formulas for the upper and lower control limits for the range chart:

$$UCL_R = D_4 \ AVER(R)$$
$$LCL_R = D_3 \ AVER(R)$$

where AVER(R) represents the Mean of the Distribution of R. This Mean is related to SD(X) by:

$$AVER(R) = d_2 \ SD(X).$$

Combining these equations, we get:

$$UCL_R = D_4 \ d_2 \ SD(X)$$
$$LCL_R = D_3 \ d_2 \ SD(X).$$

When SD(X) = Measurement Unit, these equations become

$$UCL_R = D_4 \ d_2 \text{ measurement units}$$
$$LCL_R = D_3 \ d_2 \text{ measurement units.}$$

These values are tabled for subgroup sizes of n = 2 to n = 10 in Table 1.3.

Table 1.3: Control Limits for Range Chart When SD(X) = Measurement Unit

Subgroup Size	LCL	UCL	Possible Values for Range Within Limits	Number of Possible Values for Range Within Limits
2	none	3.69	0, 1, 2, 3	4
3	none	4.36	0, 1, 2, 3, 4	5
4	none	4.70	0, 1, 2, 3, 4	5
5	none	4.92	0, 1, 2, 3, 4	5
6	none	5.08	0, 1, 2, 3, 4, 5	6
7	0.21	5.20	1, 2, 3, 4, 5	5
8	0.39	5.31	1, 2, 3, 4, 5	5
9	0.55	5.39	1, 2, 3, 4, 5	5
10	0.69	5.47	1, 2, 3, 4, 5	5

Since the values in Table 1.3 define the borderline condition, the following guidelines for detecting Inadequate Measurement Units can be established.

The measurement unit borders on being too large when there are only 5 possible values within the control limits on the Range Chart. Four values within the limits will be indicative of Inadequate Measurement Units, and fewer than four values will result in appreciable distortion of the control limits.

The only exception to this occurs when the Subgroup Size for the Range Chart is n = 2; here 4 possible values within the control limits on the Range Chart will represent the borderline condition. Three possible values within the limits will be indicative of Inadequate Measurement Units, and fewer than three values will result in appreciable distortion of the control limits.

Thus, there need be no confusion about whether or not the measurement unit being used is sufficiently small for the application at hand. The control chart clearly shows when it is not. Fortunately, when this problem exists, the solutions are straightforward. But one must implement one of these solutions before the control charts will be of any real use.

One other note is needed. Occasionally the measurement unit will be too large simply because the data are truncated to a certain level in order to avoid reporting "noise." When such truncation creates inadequate discrimination, it is also cutting off part of the signal! Recording one extra digit will usually be sufficient to eliminate this source of inadequate discrimination.

Chapter Two

Consistency and Bias

Technician One was always very careful to make sure that his test instrument was calibrated. Every hour he routinely measured and recorded the value for an item which was designated as a "reference standard." If the measurement he obtained did not match the "accepted value" for the reference standard, he would adjust his instrument to compensate for the difference between his measured value and the accepted value. Because of this hourly re-calibration of his instrument, Technician One was considered to be a very careful and conscientious worker.

Technician Two, on the second shift, used the same instrument as Technician One. He also measured the reference standard every hour, but his approach to adjustment was quite different from that of Technician One. Instead of adjusting the instrument each hour, Technician Two would plot the value obtained for the reference standard on a control chart for individual values. (The central line for this control chart was set at the accepted value for the reference standard.) Whenever this control chart indicated a lack of control, Technician Two would make an adjustment in the instrument. Otherwise, he would continue to use the instrument without making any adjustment.

These two technicians continued to operate in these different ways over a period of several months. Finally, when their supervisor became aware of the existence of these two different approaches to calibration, he decided to investigate the results obtained by the two methods. Since both technicians had recorded the test values for the reference standard, he decided to compare the histograms of these values. These histograms are shown in Figure 2.1. The scale shows the deviations from the "accepted value" for the reference standard.

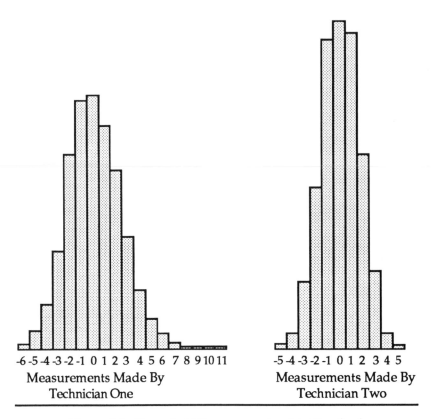

-6 -5 -4 -3 -2 -1 0 1 2 3 4 5 6 7 8 9 10 11

Measurements Made By
Technician One

-5 -4 -3 -2 -1 0 1 2 3 4 5

Measurements Made By
Technician Two

Figure 2.1: Repeated Measurements of A Standard

Technician One's histogram is wider because of his hourly adjustments. The variation of his hourly adjustments was added to the natural variation of the measurements themselves, and this increased variation is seen in the wider histogram. Many of Technician One's hourly adjustments were unnecessary, and every one of them contributed to the increased variation seen in his histogram.

Technician Two, on the other hand, has a narrower histogram because he did not make an adjustment every hour. He adjusted the process aim only when the control chart gave a clear indication of the need to adjust. In fact, Technician Two rarely made any adjustments, except at the beginning of his shift. The histograms suggest that these adjustments at the start of the shift were necessary *to undo Technician One's needless adjustments.*

Based on this study, a new calibration procedure was adopted. Control charts were made a routine part of every calibration scheme, and standard operating procedure was changed so that adjustments were to be made only in response to an indication of a lack of control. Subsequently, several test methods showed a sudden and dramatic improvement due to the removal of overcalibration.

The use of a control chart to check on the consistency of a measurement method is equivalent to the introduction of an operational definition of consistency for a measurement process.

An Operational Definition of Measurement Consistency

An operational definition consists of three parts:
- (1) a criterion to be satisfied,
- (2) a test for determining conformance to the criterion,
- (3) a decision rule for interpreting the test results.

A measurement process can be said to be consistent if repeated measurements of the same item display statistical control. This is the criterion to be satisfied. (In the case of destructive testing, one would have to test samples which are prepared to be as much alike as possible.)

The test procedure for determining conformance to this criterion is the use of a control chart for the repeated measurements. By plotting these repeated measurements on a control chart, one will be able to detect any lack of control in a timely manner.

The decision rules for interpreting the control chart of the repeated measurements will be the usual decision rules for detecting a lack of control on a control chart.

When the repeated measurements display statistical control, there is no need to re-calibrate or adjust the measurement process. The measurements are already as consistent as they can be, and any adjustments to the measurement process will merely increase the variation in the measurements.

When the repeated measurements display a lack of statistical control, there is a real need to respond by first trying to determine the reason for the inconsistency in the measurement process, and then trying to re-establish statistical control for the measurement process.

When it is the control chart for location (either an Average Chart or a Chart for Individual Values) which is out-of-control, a simple re-calibration may re-establish the measurement process in a state of statistical control. However, without a knowledge of why the measurement process went out-of-control, one either risks bad values on production samples or continually measures the reference standard in conjunction with each production sample. Knowledge of why the measurement process has gone out-of-control can often be used to avoid such problems, while simultaneously reducing the number of measurements made.

When the Range Chart is out-of-control there is a problem with the measurement process. Each out-of-control range on a chart of repeated measurements signals the presence of an unusual value. Since the actual reasons for such unusual values can vary greatly with the situation, one can only search until the Assignable Cause is found. Moreover, this will usually have to be done before one can hope to re-establish statistical control for the measurement process.

An automatic gauge is checked for consistency by placing a "reference assembly" on the conveyor belt and recording the stack height measured by the gauge. Since the dimension measured was safety-related, this check procedure was used every two hours. The stack heights are given in units of one-thousandth of an inch, with zero being fixed at the minimum specification for stack height.

The data for the reference assembly for March 28 were: 7, 9, 8, 7, 7, 8, 9, 7, 9, 7, 8, 7. The average of these 12 values is 7.75, and the average moving range is 1.2727. The XmR Chart for these 12 measurements is shown in Figure 2.2.

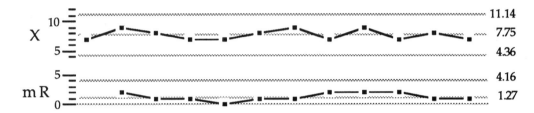

Figure 2.2: Stack Height Data for Reference Assembly on March 28

Based on the data of March 28, the standard deviation of the measurement process is estimated to be:

$$\hat{\sigma}_e = \bar{R}/d_2 = 1.2727/1.128 = 1.13 \text{ thousandths of an inch}$$

On March 29 the data for the same reference assembly were: 5, 7, 6, 4, 4, 7, 6, 4, 4, 7, 7, 4. These values are plotted against the limits for March 28 in Figure 2.3.

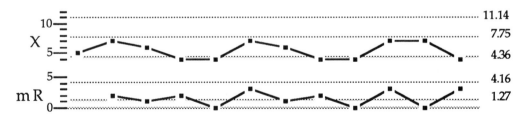

Figure 2.3: Stack Height Data for Reference Assembly on March 29

The X Chart for March 29 shows a clear and definite shift in the values obtained for the reference assembly. The Range Chart for March 29 shows no change in the dispersion of the measurements. The operator therefore looked for an assignable cause which could lower the measurements of the reference assembly. He found that the air pressure for the probe was 80 psi instead of the specified value of 50 psi. This higher air pressure caused the probe to squeeze the parts before measuring them.

With destructive testing, one must make every effort to obtain and maintain consistency from one "reference standard sample" to the next. Moreover, with sources which deteriorate with age, one will need to chart several reference standards simultaneously. By staggering the introduction of new reference standards, one can sometimes bridge across deteriorating standards and maintain a coherent record of the consistency of the measurement process.

If it is impossible or impractical to repeatedly measure reference standards, one may use duplicate measurements of production samples to check the consistency of the measurement process. The range of such duplicate measurements (or of triplicate or quadruplicate measurements) can be plotted on an ordinary Range Chart. An out-of-control range will indicate an inconsistent measurement. This simple test identifies the presence of unusual values (outliers) when duplicate measurements are used, and it helps identify the outlier itself when triplicate or quadruplicate measurements are obtained. A Range Chart which displays statistical control indicates consistency among the multiple measurements. When the measurements display this type of consistency one may estimate the standard deviation of Test-Retest Error using the Average Range.

Example 2.2:   Viscosity Measurements for Product 10F:

Viscosity values for product 10F were routinely obtained in duplicate by splitting each sample and measuring the viscosity of each half. The average of these two measurements would then be reported as the viscosity for the sample.

Since each pair of readings are duplicates, the difference between such values will represent Test-Retest Error rather than product variation. Therefore, one could plot the range of such values on an ordinary Range Chart to check for the consistency of the measurement process. The values for seven lots are shown below. These values are recorded to the nearest 10 centistokes and expressed in units of thousands of centistokes.

Lot	32	33	41	49	61	64	76
Viscosities	20.48	19.37	20.35	19.87	20.36	19.32	20.58
	20.43	19.23	20.39	19.93	20.34	19.30	20.68
R	0.05	0.14	0.04	0.06	0.02	0.02	0.10

Figure 2.4: Range Chart for Duplicate Measurements of Product 10F

The Average Range is 0.0614. The upper control limit on the Range Chart is 0.200, and no ranges are found to exceed this limit. Thus, these duplicate readings display consistency, and an estimate of the standard deviation of the measurement process is:

$$\hat{\sigma}_e = \bar{R}/d_2 = 0.0614/1.128 = 0.054 = 54 \text{ centistokes.}$$

One should not use the ranges within the subgroups in Figure 2.4 to construct control limits for the Average Viscosities. Since the subgroups represent different lots, and the ranges reflect only the Test-Retest Error, limits based upon these ranges will not make any allowance for production variation.

If one wished to use these data to track the consistency of the production process, it would be best to use the averages of these duplicate readings as if they were individual measurements. The Moving Ranges defined by these Averages would then be used to construct control limits which would reflect the variation of the production process as well as the variation due to measurement error.

Lot	32	33	41	49	61	64	76
Viscosities	20.48	19.37	20.35	19.87	20.36	19.32	20.58
	20.43	19.23	20.39	19.93	20.34	19.30	20.68
$\bar{X}$	20.455	19.300	20.370	19.900	20.350	19.310	20.630
mR		1.155	1.07	0.47	0.45	1.04	1.32

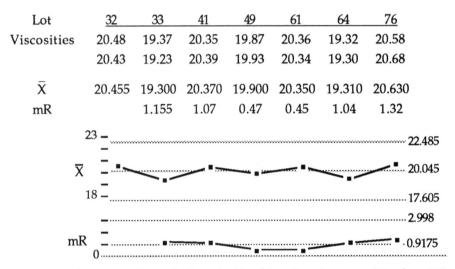

Figure 2.5: Control Chart for Tracking Consistency of Product 10F

The Grand Average Viscosity is 20.045, and the Average Moving Range is 0.9175. An estimate of SD(X) is 0.813, and control limits for the average lot viscosities are 17.605 to 22.485. The upper control limit for the moving ranges is 2.998.

Thus, the Range Chart in Figure 2.4 tracks the consistency of the viscosity measurements, while the Control Chart in Figure 2.5 tracks the consistency of the production process. These three charts are necessary because of the sources of variation present in the data and the way these sources of variation were allocated to the subgroups. The variation within the subgroups checks for consistency of the measurement process, but it does not provide an appropriate yardstick for lot-to-lot variation. If the ordinary Range Chart in Figure 2.4 goes out-of-control, look for an assignable cause in the measurement process. If the Average Chart or the Chart for Moving Ranges in Figure 2.5 goes out-of-control, then look for an assignable cause in the production process. (As always, understanding the subgrouping is the key to interpreting the charts.)

Without an operational definition of consistency for a measurement process, one will inevitably re-calibrate the measurement process either too often or too seldom. Both errors will increase the variation in the measurements, and will lead to needless adjustments of the manufacturing process. These, in turn, will create excessive variation in the production process, which will create additional costs for the manufacturer. For this reason, it is important to establish and maintain a program for checking on the consistency of those measurements which are used to operate the production process.

Checking A Measurement Procedure for Bias

While the procedures above will track measurement consistency, it is much more difficult to check for a systematic measurement bias. In fact, before one can check a given measurement procedure for bias, there must be a Master Measurement Method which can be used to establish and certify the value for a reference sample. Moreover, this Master Measurement Method must leave the reference sample unchanged. Given that these restrictions are satisfied, one may then measure the Certified Reference Sample with the usual measurement process, and plot the results on a control chart. This control chart will differ from the charts used for consistency in one respect: the central line for the chart for location will be set at the certified value of the reference sample. Thus, any indication of a lack of control on the chart for location will be an indication of bias, while a lack of control on the chart for dispersion will still indicate an inconsistency in the measurement process. If such a lack of control is found on the chart for location, the difference between the Grand Average and the certified value will estimate the bias of the measurement procedure.

Chapter Three

Interpreting Measurements

Every day, measurements are obtained and compared with one or more specification limits. As a result of such comparisons, decisions are routinely made relative to:

    (1) the measured item,

    (2) the batch of product from which the measured item was selected, and

    (3) the process which produced the measured item.

Each of these decisions represents a different use of measurements. The first usage may be termed "Descriptive," since it seeks to characterize the item measured. The other usages have been termed, respectively, as "Enumerative" and "Analytic" by Dr. W. Edwards Deming. Each usage has its own set of conditions and is justified by its own set of assumptions. When these assumptions are valid, the usage is appropriate, and one can make the decision with only a small risk of being wrong. When the assumptions for a given decision are not satisfied, the user can no longer specify or control the risk of making a mistake when he uses the measurement to make a decision.

It is the purpose of this chapter to define and clarify the distinctions between each of these three usages of measurements.

The Characterization of Measured Items: Probable Error

The first usage of measurements is the classification of the measured item relative to specifications. This could be called a "descriptive" application of the measurements. There is no extrapolation involved, simply the interpretation of that measurement relative to the product specifications. When one or more measurements are used in this manner, it will usually be helpful to know just how much uncertainty is introduced by the measurement process itself.

One way to characterize the uncertainty in a measurement is to specify the "Median Uncertainty." (Another name for this quantity is the "Probable Error" of measurement.) The Median Uncertainty is generally estimated to be $\pm\, 0.67\, \sigma_e$ where σ_e is the standard deviation for the measurement process. The terms "Probable Error" and "Median Uncertainty" will be used as synonyms.

The interpretation of Probable Error as the median uncertainty for a measurement is based upon the following argument. Consider the "uncertainty" of a given measurement to be the difference between that measurement and the grand average of hundreds of independent measurements of the same characteristic of the same part. If these "uncertainties" form an approximate normal distribution, then about one-half of the actual measurements will fall in the interval:

$$(\text{average} - 0.67\, \sigma_e \, , \, \text{average} + 0.67\, \sigma_e \,).$$

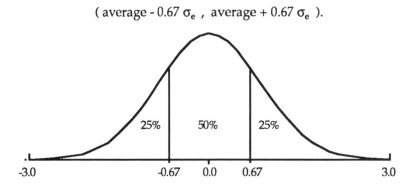

Figure 3.1: Distribution of Measurements About Their Grand Average

This means that approximately one-half of the measurements can be considered to have an "uncertainty" of less than $0.67\, \sigma_e$ while approximately one-half of the measurements can be thought to have an "uncertainty" which is greater than this amount. Thus, the Median Uncertainty for a given measurement can be taken to be $\pm\, 0.67\, \sigma_e$. This value has historically been called the Probable Error. It is a reasonable estimate of the typical uncertainty for any single measurement. It can be used to estimate the amount of leeway which should be attached to any interpretation of the measurement itself.

In essence, the Probable Error defines the effective resolution of a measurement. If a particular measurement is made to the nearest 0.001 inch, but has a Probable Error of $\pm$ 0.008 inch: the effective resolution of these measurements is 0.008 inch rather than 0.001 inch. The units of measurement generally describe the round-off involved in the measurements, while the Probable Error describes the "average" deviation between a measurement and the average of many repeated measurements of the same thing. The effective resolution of any measurement will be equal to the larger of these two quantities. When the Probable Error is smaller than the measurement unit the round-off will determine the resolution of the measurements. When the Probable Error is larger than the measurement unit the round-off will be lost in the uncertainty of the measurement.

For those situations which require a more conservative approach, the Worst Case Uncertainty is traditionally taken to be $\pm 2 \sigma_e$ (± 3 Probable Errors). From the Empirical Rule, this "Worst Case Uncertainty" can be considered to bracket approximately 95% of the "uncertainties." It should be noted that the sum of two or more Worst Case Uncertainties is meaningless. The Worst Case Uncertainty for a stack of parts will not be the sum of the Worst Case Uncertainties for the individual parts.

Both of these uncertainty figures depend upon the standard deviation of the measurement process, σ_e. Of course, such a standard deviation has meaning only as long as the measurement process is consistent (as defined in the previous chapter). Since the only way to discern if the measurement process is consistent is by maintaining one of the control charts described in Chapter Two, one should be able to use such charts to obtain an estimate of σ_e. This estimate would typically be found by dividing the Average Range for duplicate measurements by the appropriate value of d_2.

Example 3.1:   Stack Height Data:

In Example 2.1, the Automatic Stack Height Measurement System was found to have an Average Moving Range of 1.2727 thousandths of an inch. Thus, the standard deviation of the measurement process was estimated to be

$$\hat{\sigma}_e = \bar{R}/d_2 = \frac{1.2727}{1.128} = 1.13 \text{ thousandths of an inch.}$$

The Probable Error of a single measurement would be ± 0.76 thousandths, and a Worst Case Uncertainty would be ± 2.3 thousandths. Thus, the Stack Height for a given assembly can be taken to be approximately correct to the nearest 1 thousandth of an inch, and at most, in error by about 2 thousandths of an inch. Moreover, since the measurement unit is 1 one-thousandth of an inch, the effective resolution of these measurements is essentially 1 one-thousandth of an inch.

Example 3.2:   Viscosity Measurements:

In Example 2.2, the viscosity measurements for product 10F were found to have a standard deviation of 54 centistokes. Thus, given a small volume of product 10F which has been thoroughly mixed, and given a measurement of the viscosity of this small volume, the viscosity essentially is known to within ± 36 centistokes. At most, the viscosity of the small volume is unlikely to differ from the measured value by more than ± 109 cs. The effective resolution of these viscosity measurements is given by the Probable Error, ± 36 centistokes. (The measurement unit is 10 centistokes.)

The Classification of Measured Items Relative to Specifications

When measurements are used for 100 percent screening of product, the problem of how to compare the measurements with the specification limits must be addressed. How can one make an allowance for measurement error, and how large an allowance is needed? One approach uses specification limits which have been "adjusted" by multiples of the Probable Error.

A measured item may be judged to be conforming when the measurement falls within the "PE Tightened Specification Limits." These PE Tightened Specification Limits are obtained by adjusting each of the Specification Limits toward each other by an amount equal to the Probable Error.

Likewise, a measured item may be judged to be nonconforming when the measurement falls outside the "PE Widened Specification Limits." These widened limits are obtained by adjusting the Specification Limits away from each other by an amount equal to the Probable Error.

If the measurement falls inside the PE Widened Limits and outside the PE Tightened Limits, then the measured item may be judged to be a borderline item. A clear resolution of the conformity of such borderline items will require additional measurements of the item. When such measurements are not worthwhile, the supplier and the customer may be able to negotiate the disposition of borderline items.

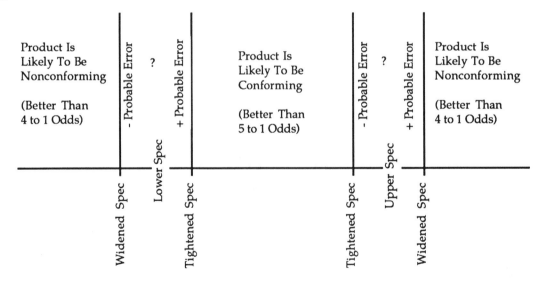

Figure 3.2: PE Widened and Tightened Specification Limits

If (n-1) additional measurements are obtained for a borderline item, the n values should be averaged together and their average should be compared to specification limits which have been widened and tightened by the Probable Error of the average of n measurements. Averages which remain in the region of uncertainty will still indicate borderline items, averages within the PE Tightened Limits will indicate a conforming item, and averages outside the PE Widened Limits will indicate a nonconforming item.

Assuming that an item is equally likely to be either conforming or nonconforming, the procedures above will correctly classify product at least 80 percent of the time. (This probability was computed using Bayes' Theorem and the *a priori* assumption of equally likely states.)

If the consequences of a misclassification are severe, this 80 percent chance of correctly classifying product may not be acceptable. When this happens, the procedures may be modified in order to reduce the chance of misclassification, but only at the expense of obtaining more borderline items. Define the 2 PE Widened and 2 PE Tightened Specification Limits to be set at a distance equal to twice the Probable Error on either side of the Specification Limits. With these limits the chances of a correct classification go up to around 95 percent for those items with measurements which are clearly within the tightened limits or outside the widened limits. Unfortunately, there may be an increased number of items with measurements in the borderline category.

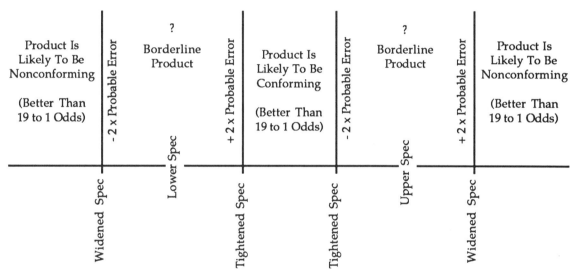

Figure 3.3: 2 PE Widened and 2 PE Tightened Specification Limits

<u>Example 3.3:</u> <u>Stack Height Data:</u>

In Example 3.1, the Automatic Stack Height Measurement System was found to have a Median Uncertainty of ± 0.76 thousandths (± 0.00076 inch). If the specifications for Stack Height are 4.000 in. to 4.005 in., then the PE Tightened Specification Limits will be 4.00076 in. to 4.00424 in., while the PE Widened Limits will be 3.99924 in. to 4.00576 in. If the measurements are made to the nearest 0.001 in., then values of 3.999 and below indicate nonconforming items, values of 4.006 and above indicate nonconforming items, and values of 4.001 to 4.004 indicate conforming items. In each case, the odds of a correct classification are better than 4 to 1. A value of 4.000, or a value of 4.005, will indicate a borderline item.

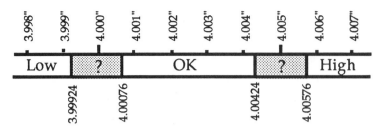

Figure 3.4: Specification Limits Adjusted by the Probable Error

With severe consequences of misclassification, one could increase the odds of a correct classification as follows: the 2 PE Tightened Specification Limits equal 4.00152 in. to 4.00348 in., and the 2 PE Widened Specification Limits equal 3.99848 in. to 4.00652 in. Using these limits the values of 4.002 in. and 4.003 in. indicate conforming items, while values of 3.998 in. or smaller, or 4.007 in. or larger, indicate nonconforming items. Here the odds of correct classification are better than 19 to 1. However, values of 3.999 in., 4.000 in., 4.001 in., 4.004 in., 4.005 in., and 4.006 in. will now indicate borderline items.

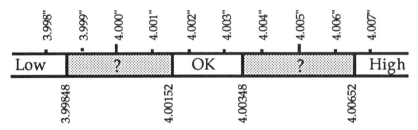

Figure 3.5: Specification Limits Adjusted by Twice the Probable Error

Thus, while one can increase the odds of correct classification for some items by using 2 PE Widened and 2 PE Tightened Specification Limits, one will simultaneously increase the chances of obtaining product in the borderline category. This will usually be undesirable. Of course, if the production process displays statistical control, then the relationship between the Natural Process Limits and the Adjusted Specification Limits will determine if this increased risk of obtaining borderline items is appreciable.

Finally, if the situation warrants, one could define 3 PE Widened and 3 PE Tightened Specification Limits by using ± 3 Probable Errors. Under the assumption that conformity and nonconformity are equally likely, a measurement within the 3 PE Tightened Specification Limits would indicate a conforming item with better than 99 to 1 odds of a correct classification, and a measurement outside the 3 PE Widened Specification Limits would indicate a nonconforming item with better than 99 to 1 odds of a correct classification. Once again, while some items may be more certainly classified, the region of uncertainty will be much larger, and the distance between definitely nonconforming product and definitely conforming product will be larger than with the other schemes.

Of course the preceding intervals could be combined into a stair-step approach if so desired. Product outside the 3 PE Widened Limits will be nonconforming with odds of better than 99 to 1. Product within the 3 PE Widened Limits and outside the 2 PE Widened Limits will be nonconforming with odds of 19 to 1. Product within the 2 PE Widened Limits and outside the PE Widened Limits will be nonconforming with odds of 4 to 1. Product within the PE Widened Limits and outside the Specification Limits will be nonconforming with odds of 3 to 2. Product within the Specification Limits and outside the PE Tightened Limits will be conforming with odds of 2 to 1. Product within the PE Tightened Limits and outside the 2 PE Tightened Limits will be conforming with odds of better than 5 to 1. Product within the 2 PE Tightened Limits and outside the 3 PE Tightened Limits will be conforming with odds of better than 19 to 1. Product within the 3 PE Tightened Limits will be conforming with odds of better than 99 to 1. This stair-step approach is summarized in Figure 3.6.

```
                                                      1 in 100  odds conforming
                            3 PE Widened Specification Limit
                                                5 in 100  odds conforming
                        2 PE Widened Specification Limit
                                            19 in 100  odds conforming
                    PE Widened Specification Limit
                                        39 in 100 odds conforming
                Upper Specification Limit
                                    64 in 100 odds conforming
            PE Tightened Specification Limit
                                84 in 100 odds conforming
        2 PE Tightened Specification Limit
                            95 in 100 odds conforming
    3 PE Tightened Specification Limit
                99 in 100 odds conforming
    3 PE Tightened Specification Limit
                            95 in 100  odds conforming
        2 PE Tightened Specification Limit
                                84 in 100 odds conforming
            PE Tightened Specification Limit
                                    64 in 100 odds conforming
                Lower Specification Limit
                                        39 in 100 odds conforming
                    PE Widened Specification Limit
                                            19 in 100 odds conforming
                        2 PE Widened Specification Limit
                                                5 in 100 odds conforming
                            3 PE Widened Specification Limit
                                                      1 in 100 odds conforming
```

Figure 3.6: The Odds of Conforming Product

A proper understanding of how much uncertainty is attached to any single measurement is crucial to a proper use of measurements. Without such an understanding, one will be misled by the apparent objectivity of the measurements and, in consequence, may make incorrect decisions. The approaches outlined above are intended to clarify just what is known, and also what is not known, when using a measurement to characterize the measured item.

Many corporate manuals characterize measurements by comparing the standard deviation of Test-Retest Error with Specification Limits. Generally this is done by condemning the measurement process if the standard deviation of Test-Retest Error exceeds some fixed percentage (commonly 20%) of the Specified Tolerance. This approach essentially sets a specification on how good the measurements should be, with the implication that one should purchase a better measurement system whenever the current system fails to meet this specification.

Moreover, most corporate manuals simply give this guideline *ex cathedra*, making no attempt to either justify or explain it. As a result, the user is commonly left without any interpretative guide on just what can be done with the measurements. If the measurement process is judged to be bad, they generally attempt to find a better measurement process, even though this expense may not be justified. The fallacy of this approach is that it does not consider the very important question of whether or not the measurement can detect variation in the product. This issue will be considered more completely at the end of Chapter Five.

Using the approach of adjusting the Specification Limits with the Probable Error has the advantage of spelling out exactly what can be accomplished with the measurements in different situations. For example, the "99 in 100 odds conforming category" from Figure 3.6 will be obscured by the PE Adjusted Limits only when the standard deviation of Test-Retest Error exceeds 25% of the Specified Tolerance. The "95 in 100 odds conforming category" will be obscured by the PE Adjusted Limits only when the standard deviation for Test-Retest Error exceeds 37.5% of the Specified Tolerance. Finally, the "84 in 100 odds conforming category" will be obscured only when the standard deviation of Test-Retest Error exceeds 75% of the Specified Tolerance. Thus, while none of these situations are as desirable as having a very small standard deviation for Test-Retest Error, the measurements still have some utility even when they "flunk" the "20% of Specified Tolerance" guideline.

It might be instructive to consider Example 3.3 in the light of the "20% of Specified Tolerance" guideline. The standard deviation of Test-Retest Error for the Automated Stack-Height Measurement System was 1.13 thousandths, and the Specified Tolerance for Stack-Height was 5 thousandths. Thus, this measurement process fails this guideline since the standard deviation for Test-Retest Error is 22.6% of the Specified Tolerance. However, as can be seen in both Figures 3.4 and 3.5, certain measurement values do strongly indicate conformity to specifications, while other values strongly indicate nonconformity to specifications.

The Characterization of Batches of Product

The use of measurements to characterize product which has not been measured introduces another source of uncertainty which the user must consider. In addition to the uncertainty of the measurement process itself, there is the uncertainty of the extrapolation from the measured item or items to the unmeasured product. This uncertainty of extrapolation is usually much greater than the uncertainty of the measurement process. In some cases this uncertainty of extrapolation can be quantified. In other cases one can only cross one's fingers and hope that the uncertainty of extrapolation will not lead to the wrong decision.

Just what does it take to quantify the uncertainty of extrapolation when characterizing product with one or more measurements? This question is the basic question of sampling theory. First, there must be a certain rationality to the extrapolation: the measured item or items must come from the batch of product which is being characterized. Next, there must be some basis for thinking that the sample (the measured product) is representative of the batch (the unmeasured product). This is usually justified on the basis of the sampling procedure. One common feature of sampling procedures is the existence of a sampling frame. The frame is simply the collection of material (product) from which the sample was obtained. One should note that the frame may differ (slightly) from the batch of product being characterized. As long as the frame approximates the batch of product, the rationality of the extrapolation is preserved. If the difference between the frame and the batch of product becomes very great, the rationality of the extrapolation will be destroyed. Finally, in addition to the frame, the sampling procedure must specify how items are to be selected from the frame in such a way that every item in the frame has an equal chance of being chosen (Dr. Deming's "equal and complete coverage"). When these conditions are satisfied, the measurements may be used to extrapolate from the product measured to that product which has not been measured, and the techniques of enumerative statistics may be used to quantify the uncertainty of the extrapolation.

The specification of the sampling frame, and the specification of the selection procedure are both complex and time-consuming tasks. With expensive one-time studies, the success of the study will frequently depend upon the careful definition of frame and sampling procedure. However, in routine industrial applications it is uncommon to find a well-defined sampling frame. In some of the better situations one occasionally finds a standard sampling procedure for a given product. But in most industrial situations, there is neither a sampling frame nor a sampling plan that gives equal and complete coverage. Therefore, given the lack of these prerequisites in industrial applications, **it will almost always be impossible to quantify the uncertainties associated with the extrapolation from the measured sample to the batch of product.** When this happens, there will be no basis for using statistical inference, and there will be no way to know how reasonable the extrapolation from measured sample to batch of product may be. At best, the samples can be characterized as convenience samples or haphazard samples, with the implicit connotation that all extrapolation based on these samples is hazardous.

The problem with haphazard or convenience samples is the possibility of subjective bias in the selection of the sample. Any subjectivity in sample selection will undermine the hoped for "representativeness" of the sample, and destroy the rationality of the extrapolation.

Given that most industrial samples are convenience samples, how can one ever hope to use the measurements to characterize the product in a batch, or to characterize the product stream? The honest answer is that, unless the production process displays statistical control, one cannot. The existence of a state of statistical control will provide a basis for using sample measurements to characterize the product stream. When a process displays statistical control, it is predictable and consistent. The predictability will justify the extrapolation from the sample to the product not measured, and the consistency justifies the assumption that the sample is "representative" of the product stream.

At the same time, the existence of a state of statistical control will tend to make it less imperative that each batch of product be characterized as conforming or nonconforming. As long as the process displays statistical control, the product stream must be considered homogeneous. When a process displays statistical control, there will be no indications that the product stream is changing over time, and the classification of some portions of that stream as "conforming batches" and other portions as "nonconforming batches" is both arbitrary and capricious.

Therefore, when each "lot" or "batch" in a product stream is characterized by the measurements obtained for a sample drawn from that lot or batch, one should plot the sequence of measurements on a control chart.

If this control chart displays statistical control, then one is reassured that the product is consistent "batch" to "batch", and all the "batches" should be treated alike. That is, the product stream must be considered homogeneous, and all the product should be treated alike. Either the product stream contains some nonconforming items, and every single item must be measured and judged relative to specs, or else the product stream should be considered to consist of 100% conforming product, and individual items need only be checked for the purposes of maintaining the control chart.

When the control chart of sample measurements does not display statistical control, one may attempt to subdivide the product stream into "conforming batches" and "nonconforming batches." However, the very unpredictability of an out-of-control process will usually make this subdivision imperfect. If the process is changing quickly relative to the sample frequency, there is little hope that any single sample measurement is representative of the product produced shortly before or after the sampled product, and the subdivision of the product stream will be arbitrary and ineffective.

On the other hand, if the process is one which goes out of control slowly, relative to the sample frequency, then this slow change reassures the user that a sample measurement will characterize the product produced shortly before and after the sampled product, and therefore the product stream may be rationally subdivided into "conforming batches" and "nonconforming batches."

Example 3.4: _____ Percent Solids Data:

A certain bulk product is produced by a continuous process. Every four hours a sample is taken from the process outflow and analyzed for percent solids. Each of these measurements is used to characterize the product produced during the previous four-hour period. The measured percent solids for the past 100 samples are recorded in columns below, along with the moving ranges.

Table 3.1 Percent Solids Data

X	mR	X	mR	X	mR	X	mR	X	mR	X	mR	X	mR
67		68	0	71	1	69	0	67	1	67	3	72	2
69	2	70	2	70	1	71	2	69	2	69	2	67	5
68	1	67	3	72	2	70	1	68	1	67	2	68	1
71	3	65	2	69	3	69	1	66	2	67	0	68	0
69	2	69	4	68	1	71	2	67	1	72	5	67	1
69	0	67	2	66	2	71	0	68	1	69	3	71	4
70	1	70	3	67	1	66	5	71	3	70	1	71	0
68	2	71	1	68	1	72	6	68	3	72	2	69	2
68	0	68	3	70	2	68	4	70	2	69	3	70	1
70	2	69	1	70	0	70	2	71	1	64	5	70	0
69	1	68	1	72	2	74	4	66	5	67	3		
72	3	69	1	69	3	72	2	71	5	69	2		
66	6	68	1	71	2	71	1	68	3	68	1		
69	3	68	0	68	3	71	0	71	3	70	2		
68	1	70	2	69	1	68	3	70	1	70	0		

The Average Measured Percent Solids for these 100 samples is 69.07%. The Average Moving Range for these data is 1.97%. Based on these statistics, the Upper Natural Process Limit is estimated to be 74.31%, the Lower Natural Process Limit is estimated to be 63.83%, and the Upper Control Limit for the moving ranges is estimated to be 6.4%. The XmR Chart for these data is shown below.

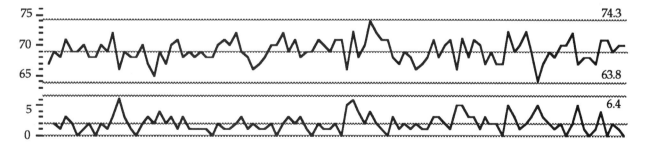

Figure 3.7: XmR Chart for Percent Solids Data

These measurements display statistical control. There is absolutely no indication of any instability or process drift in these data. Therefore, one must conclude that the product produced by this process is as consistent as this process is currently capable of producing. Product consistency can be improved only by changing the process in some fundamental way.

The Specification Limits are 66.5% to 71.5%. Since these Specification Limits fall within the Natural Process Limits it is inevitable that some measurements will fall outside of the Specification Limits even while the process remains in control and the product stream remains consistent. What should be done when a measurement falls outside the Specification Limits but within the Natural Process Limits?

First, one should observe that the grand average of 69.07% is very close to the mid-spec point of 69%. Thus, the process is almost perfectly centered relative to the Specification Limits. Next, one should observe that, at the sample frequency of this chart, there is no evidence of positive autocorrelation between successive points. This suggests that the percent solids can change considerably over a four-hour period. (Positive autocorrelation exists when successive values are quite similar. If the process was changing slowly, relative to the sample frequency, then one would expect to see positively autocorrelated data.) Therefore, as long as the process remains in control, it would be a mistake to quarantine a batch of product based upon a nonconforming measurement.

The notion that one can accept some batches, and reject others, based upon the value for a sample drawn from each shipment, is equivalent to playing roulette with the product. The usual assumption used to justify such acceptance sampling is that the lot quality is highly variable from batch to batch. If this is the case, then how can one be assured regarding the homogeneity of the product within each batch? The classification of each batch as conforming or nonconforming will depend upon the extrapolation from the product measured to that product which has not been measured, and the assumption of homogeneity within the batch will be essential for this extrapolation to make sense. So using acceptance sampling is rather like wanting to eat your cake and have it too: one must assume that the product quality is highly variable from batch to batch, but is very uniform within each batch. If this rather unusual combination of conditions does not apply, then one will have either a poor basis for extrapolation, or will be making arbitrary judgements. As many have discovered the hard way, you cannot inspect quality into the product stream.

If the product batches are true batches, and the conditions above apply, then acceptance sampling may make some sense. On the other hand, if the "batches" are arbitrary segments out of the production stream, then one can rationalize the use of an in-control-but-nonconforming measurement to characterize a batch of product as nonconforming if and only if the control chart shows the process to be changing slowly relative to the sampling frequency. (When a process is changing slowly relative to the sample frequency the data will display a definite positive autocorrelation.)

Throughout this approach no explicit allowance has been made for measurement error. Such an allowance was unnecessary since the control chart already makes it. Moreover, since any attempt to use sample measurements to characterize product which has not been measured should always be based upon a control chart, the question of allowances for measurement error is moot. The real question concerns the interpretation of the measurements in the context of the control chart and what it reveals about the production process. Finally, the failure to use a control chart in this situation will inevitably result in arbitrary and capricious decisions regarding product.

The Decision Regarding Process Adjustment

Finally, the third usage of measurements is that of making adjustments to the production process. This usage involves an extrapolation from the sample measurement back to the underlying process. Dr. Deming has called such extrapolations "analytic uses" of data.

Typically, where control charts have not been utilized, this usage has been carried out along the following lines: a sample is obtained and measured; if the measurement falls below the Lower Specification Limit then the process aim is adjusted upward, and if the measurement falls above the Upper Specification Limit then the process aim is adjusted downward. This procedure will be illustrated using the Percent Solids Data.

Assume that the process aim is adjusted every time the Measured Percent Solids falls outside the specification range. In Table 3.2 the out-of-spec values have asterisks attached. The amount of each adjustment is shown in parentheses beside each out-of-spec value. Thus, Table 3.2 starts out with the values shown in Table 3.1, however, beginning with the first adjustment and continuing throughout the remainder of the data, the values are shifted by an amount equal to the cumulative adjustments to that point. Thus, the data below represent what might have been found if this adjustment procedure had been used in practice. (Read Table 3.2 in columns.)

Table 3.2: Adjusted Percent Solids Data

X	mR	X	mR	X	mR	X	mR	X	mR
67		67	6	71	2	68	2	66*(+3)	8
69	2	70	3	68	3	70	2	70	4
68	1	71	1	70	2	69	1	72*(-3)	2
71	3	68	3	67	3	67	2	66*(+3)	6
69	2	69	1	68	1	68	1	64*(+5)	2
69	0	68	1	68	0	69	1	72*(-3)	8
70	1	69	1	70	2	72*(-3)	3	71	1
68	2	68	1	69	1	66*(+3)	6	70	1
68	0	68	0	68	1	71	5	72*(-3)	2
70	2	70	2	70	2	72*(-3)	1	69	3
69	1	71	1	70	0	64*(+5)	8	71	2
72*(-3)	3	70	1	65*(+4)	5	74*(-5)	10	66*(+3)	5
63*(+6)	9	72*(-3)	2	75*(-6)	10	66*(+3)	8	70	4
72*(-3)	9	66*(+3)	6	65*(+4)	10	72*(-3)	6	70	0
68	4	68	2	71	6	68	4	69	1
68	0	66*(+3)	2	75*(-6)	4	65*(+4)	3	73*(-4)	4
70	2	70	4	67	8	71	6	69	4
67	3	71	1	66*(+3)	1	69	2	67	2
65*(+4)	2	73*(-4)	2	69	3	69	0	68	1
73*(-4)	8	69	4	66*(+3)	3	74*(-5)	5	68	0

The data of Table 3.2 are summarized in Figures 3.8 to 3.11. In Figures 3.8 to 3.10 the gray band defines the region within the Specification Limits of 66.5 to 71.5. The central line in these figures is the target value of 69. Figure 3.8 shows the original (unadjusted) Percent Solids Data. Figure 3.9 shows the effect of the cumulative adjustments upon the process aim. Figure 3.10 shows the sum of Figures 3.8 and 3.9: the Adjusted Percent Solids Values. Finally, Figure 3.11 compares the histograms of the original Percent Solids Data with that of the Adjusted Percent Solids Data.

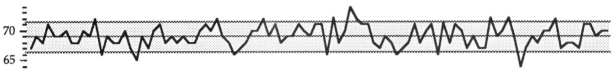

Figure 3.8: Original Percent Solids Data Relative to Specification Limits

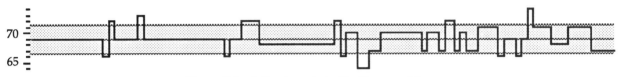

Figure 3.9: Adjustments to the Process Aim

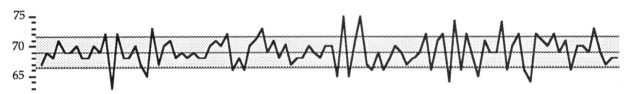

Figure 3.10: Adjusted Percent Solids Data Relative to Specification Limits

The original time series was aimed at a value of 69 throughout the time period covered by these data. The adjusted time series was aimed at 69 for only 34% of the time. It was aimed at 64 for 2% of the time, at 66 for 4% of the time, at 67 for 12% of the time, at 68 for 16% of the time, at 70 for 12% of the time, at 71 for 11% of the time, at 72 for 6% of the time, at 73 for 1% of the time and at 74 for 1% of the time. Thus, the Process Aim was outside of the Specification Limits 15% of the time because of the adjustment procedure used!

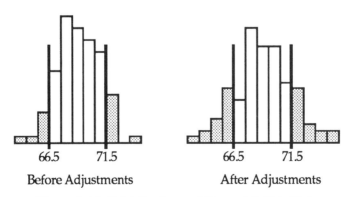

Before Adjustments   After Adjustments

Figure 3.11: Distributions of Percent Solids Data

While there were initially 16 values outside the Specification Limits, the net effect of all the adjustments was a total of 32 values outside the Specification Limits. Thus, this adjustment procedure changed a process that produced 16% nonconforming product into one that produced 32% nonconforming product! Such magic is found in all types of manufacturing operations.

The preceding example should serve to illustrate the fallacy of adjusting an incapable process based upon sample values falling outside the Specification Limits. **The failure to use a control chart as the basis for adjusting a production process is one of the best ways known to man to increase both product variation and production costs!** One will inevitably either overadjust or underadjust the process. Either way, the result will always be greater product variation and greater costs. A true theorem which could be formulated at this point is:

If a process does not display a lack of control, then it is a definite mistake to make an adjustment to the process. [*]

There is no short cut around this theorem. It looms up every time one attempts to interpret data coming from a production process. The sooner one learns this theorem, the sooner one can begin to make better product.

Limitations on Detecting Process Capability

The Test-Retest Error places some limitations on just how good a production process may appear to be. In particular, the standard deviation of product measurements can never be less than the standard deviation of Test-Retest Error.

This fact places a bound upon the various capability indexes and ratios. For example, if the Capability Ratio is defined to be

$$\frac{\text{Upper Spec - Lower Spec}}{6 \text{ Est. SD(X)}}$$

then, since Est. SD(X) $\geq \sigma_e$ must always be true, the Capability Ratio can never exceed

$$\frac{\text{Upper Spec - Lower Spec}}{6\, \sigma_e}.$$

Many customers have unknowingly demanded that their suppliers do better than this bound will allow.

Likewise, except for aberrations caused by Inadequate Measurement Units, the Natural Process Spread can never be less than $6\, \sigma_e$. As the Natural Process Spread approaches this limit measurement error becomes the dominant portion of the product measurements, and the ability to detect product

---

[*] Of course there is much more to the effective use of a control chart than simply making an adjustment. Until one knows why the process has gone out of control, one cannot know what to adjust. Moreover, the more powerful use of control charts is the discovery and removal of assignable causes of uncontrolled variation. Adjustments made in ignorance of the assignable cause may well make things worse.

variation within the Natural Process Limits is lost.

When the process capability approaches the bound above, or when the Natural Process Spread approaches $6\sigma_e$, it will generally be quite difficult to isolate the product standard deviation from the standard deviation of the measurements. This is due to the fact that the product standard deviation can be found only by subtracting the estimated variance for Test-Retest Error from the estimated variance of the measurements. Since the conditions cited above indicate that these two variances will be quite similar, their difference will be quite small. The uncertainty of the two estimated variances will generally overwhelm the small difference, so that the value obtained for the estimate of the product standard deviation will be quite unreliable.

Summary

When measurements are used in a descriptive mode, to characterize the item measured, one may use the Probable Error to make allowance for the effects of measurement error.

When measurements are used in an enumerative mode, to characterize product which has not been measured, one should first place the measurements on a control chart. The control chart will then provide a context for interpreting the measurements consistently and properly. Without this context, the extrapolation from the sample measurement to the product not measured will be quite unreliable. When using a control chart, it is unnecessary to make any adjustments for measurement error because this is done automatically by the control chart.

When measurements are used in an analytic mode, to characterize the underlying production process, one must place them on a control chart. The control chart is the only tool that makes the underlying production process visible. Adjustments to the process should be undertaken only when the control chart displays a lack of statistical control.

On the other hand, it is only when the process displays statistical control that one is able to quickly and easily assess the impact of process modifications. Thus, operations which achieve and maintain a state of statistical control are the foundation for continual improvement. Without the consistency that comes from statistical control, process modifications will always fail to live up to their expectations. Thus, while other techniques besides a control chart may be used to indicate when to make process adjustments, none of these other techniques are designed to lead to that consistency of operations which is the foundation of continual improvement. It is this potential to provide such consistency which is the real strength and power of Shewhart's control charts. While much more needs to be said about this, it is unfortunately beyond the scope of this book.

Chapter Four

EMP Studies:

Identifying Components of Measurement Error

Measurement Error typically consists of several different components. In addition to the basic component of Test-Retest Error, some obvious potential sources of measurement variation are Operator Effects, Instrument Effects, Laboratory Effects, and Day-to-Day Effects. While the basic control chart for tracking measurement consistency can provide an estimate of Test-Retest Error, a special study is required to isolate and identify these additional sources of measurement error. Since studies of this type constitute an evaluation of the measurement process, they will be referred to as EMP Studies. The way to organize, conduct and analyze such studies is outlined in this chapter.

Each of the different Components of Measurement Error can affect the measurements either by shifting them a fixed amount or by changing the variation from measurement to measurement. When the measurements are shifted a fixed amount it is called a **Bias Effect**. An **Operator Bias** exists when different operators get detectably different average values for the same thing. A **Machine Bias** exists when different machines yield detectably different average values for the same thing.

When the variation of the measurement process changes due to the presence of some factor, it is called an **Inconsistency Effect**. When different operators display detectably different Test-Retest Errors, or when some operators repeatedly display out-of-control ranges, then the measurement process can be said to be changing with the different operators, and this effect is denoted as an **Operator Inconsistency**. If the Test-Retest Error changes from one machine to another, then the measurements display a **Machine Inconsistency**.

Of course, other factors, such as laboratories and day of test, can display biases and/or inconsistencies. Regardless of the source of variation, it will always be desirable to eliminate any and all such biases and inconsistencies from the measurement process. But before this can occur, one must be able to detect and quantify these biases and inconsistencies.

This is why it is not satisfactory to merely identify the presence of these different components of measurement error. If one is to eliminate biases, one must know about each bias. If one is to eliminate inconsistencies, one must know exactly where they exist. Finally, if it proves to be impossible to eliminate a particular bias or inconsistency, then one will have to make allowances for this source of variation in the measurement process. General knowledge that certain Components of Variance exist will be insufficient. Specific knowledge will always be required. This means that the traditional "random-effect" Component of Variance models, taught in many statistics courses, will be less than satisfactory in industrial studies of the measurement process.

Since a given measurement process is unlikely to display all of these Components of Measurement Error, the first step in improving any measurement process will be the identification of just which components are present.

The Basic EMP Study for
Evaluating the Measurement Process

In order to quantify the Components of Measurement Error one must obtain measurements in such a way that each Component of Measurement Error is clearly identifiable. This means that if one is to compare three operators, each of the operators will have to be studied under conditions which are as nearly identical as possible. If one is to compare two instruments, each of the instruments will have to be studied under conditions which are as nearly identical as possible. The Basic EMP Study will involve collecting multiple measurements of the same product under conditions which permit comparisons between operators, machines, laboratories, or other relevant factors.

The Basic EMP Study will be illustrated using data obtained from a simple measurement system which uses a jig and a dial gauge to measure the length of a particular part. The whole measurement system will be referred to as "Gauge 109." Since this measurement system involves only one instrument, there can be no machine-to-machine or lab-to-lab variation. However, different operators could use the measurement system differently, so there is the potential for operator-to-operator differences. Therefore, the Basic EMP Study is designed to check for Operator Effects.

Three operators participated in this EMP Study of Gauge 109. Five parts were used in order to introduce some product variation into the EMP Study. In order to avoid confusion during the course of the study, these five parts were numbered. The actual study consisted of having each of the three operators measure each of the five parts twice. In order to keep the measurements as independent as possible, the study was conducted in the following manner: Operator A measured each of the five parts once, then Operator B measured each of the five parts once, followed by Operator C; when everyone had measured the five parts once, they then repeated the whole process, giving a total of 30 measurements.

These 30 measurements were arranged into 15 subgroups of size 2 and placed on an ordinary Average and Range Control Chart, with the two measurements in each subgroup consisting of the repeated measurements of the same part by the same operator. The sources of variation present in these data are (1) Test-Retest variation, (2) Operator-to-Operator variation, and (3) Part-to-Part variation. The Test-Retest variation shows up within each subgroup, while the Operator-to-Operator differences and Part-to-Part differences show up between subgroups. Thus, the Range Chart provides a check on the consistency of the measurement process (test-retest variation), while the Average Chart provides both a check on Operator-to-Operator consistency and a relative measure of the usefulness of the measurements for distinguishing parts. The data and the control chart are shown in Example 4.1.

<u>Example 4.1:</u>   <u>Basic EMP Study for Gauge 109:</u>

Gauge 109 is a measuring jig used to measure a certain dimension on product 109. The instrument used with this jig is marked in increments of 0.001 mm. The data are recorded in units of 0.001 mm and a value of 0 is equal to a dimension of 35.200 mm. Three operators participated in this study, and five parts were used. These five parts were selected from the product stream on each of five consecutive days. Each of these five parts was measured twice by each of the three operators. The 30 measurements of these five parts were then arranged into 15 subgroups of size 2 and placed on a control chart.

Oper.	A	A	A	A	A	B	B	B	B	B	C	C	C	C	C
Part	1	2	3	4	5	1	2	3	4	5	1	2	3	4	5
Values	34	56	6	50	33	43	49	17	54	24	35	46	10	51	25
	45	44	19	55	17	32	37	5	54	18	26	43	16	55	11
Averages	39.5	50	12.5	52.5	25	37.5	43	11	54	21	30.5	44.5	13	53	18
Ranges	11	12	13	5	16	11	12	12	0	6	9	3	6	4	14

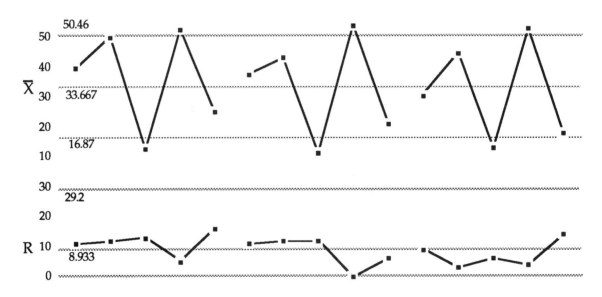

Figure 4.1: Control Chart for Basic EMP Study: Gauge 109 Data

The Range Chart displays statistical control, and the measurement unit is small enough to properly reflect the variation within the subgroups, therefore the standard deviation for Test-Retest Error is estimated to be

$$\hat{\sigma}_e \ = \ \bar{R} / d_2 \quad 8.933 / 1.128 \ = \ 7.92 \, \text{units}$$

giving a Probable Error of ± 5.3 units. Therefore, the effective resolution of these measurements is actually 5 thousandths of a millimeter rather than the stated value of 0.001 mm.

The purpose of the preceding study was the evaluation of a new measurement system. In addition to the Part-to-Part variation, this study examined three potential Components of Measurement Error: (1) the basic component of Test-Retest Error, (2) possible Operator Bias and (3) possible Operator Inconsistency. In order to make these potential Operator Effects identifiable, the data were arranged to isolate the Test-Retest Error component within the subgroups, while making any Operator Effects show up between the subgroups on either the Average Chart or the Range Chart.

Since the Test-Retest Error component is isolated within the subgroups, it is the only source of variation involved in the computation of the control limits. This fact is the key to the interpretation of the control charts.

The first consequence of this subgrouping is that the Range Chart in Figure 4.1 checks the consistency of the measurement process. When all the subgroup ranges fall within the control limits the measurement process is considered to be stable and consistent. On the other hand, any out-of-control range point should be investigated. The Assignable Cause might be operator specific, or simply that the measurement process occasionally goes out of control. But whatever the Assignable Cause, an out-of-control range will always indicate some inconsistency in the measurement process.

A second consequence of this subgrouping is that the control limits will be computed using the Test-Retest Error. Since the Test-Retest Error does not make any allowance for the Part-to-Part variation, the Average Chart should appear to be out-of-control. In fact, the more "out-of-control" the Average Chart appears to be, the better the measurement process. The control limits on the Average Chart effectively define that amount of product variation which is obscured due to Test-Retest Error. If this amount of variation swamps the Part-to-Part variation, then either the parts are unusually uniform, or else the measurements are mostly noise.

Comparing the visual band swept out by the three running records in the Average Chart of Figure 4.1 with the width of the control chart limits, it would not appear that the Part-to-Part variation is much greater than the Test-Retest Error variation. If this is the case, it will be difficult to use the Gauge 109 measurements to discriminate between parts. A way to quantify the relative usefulness of these measurements will be given in Chapter Five.

The technique used above may be extended to many different situations. In each case, the objective is to isolate Test-Retest Error within the subgroups and to allocate every other source of variation to appear between subgroups. When this is done, the Average Range can be used to estimate

the standard deviation for Test-Retest Error, the Range Chart will check for consistency of Test-Retest Error, and the Average Chart will show up other potential sources of variation.

If the sample parts or sample batches have been selected in a way that adequately reflects the natural variation for a given product, then the Average Chart will also display the relative usefulness of the measurements for that product. In general, when there is Product Variation present in the study, the narrower the control limits are relative to the band swept out by the running records, the more useful the measurements will be relative to that particular product.

One simple way to ensure that the EMP Study contains a reasonable amount of Part-to-Part variation is to select one sample part or batch on each of several successive days of operation. When this is done, the sample parts or batches are unlikely to be overly homogeneous, and the Average Chart will properly reflect the relative usefulness of the measurement process for that product.

If one deliberately selects samples based on "known" values, then the Average Chart will have to be interpreted differently than when the samples are haphazardly selected from the product stream over an extended time period.

When samples are haphazardly selected from the product stream, they are considered to reflect the product variation. When the samples are deliberately selected so that one sample will yield a "high" measurement and another sample will yield a "low" reading, the samples cannot be considered to reflect product or process variation. Instead such samples will represent some specific range of measurement values. In this latter case, the Average Chart will reflect the usefulness of the measurement process over this range of interest, rather than the usefulness relative to a specific product.

The data in Basic EMP Studies, such as the Gauge 109 Study, can also be used to obtain additional control charts which will check for detectable Bias Effects and detectable Inconsistency Effects.

EMP Studies for Bias Effects

A Bias Effect exists when some identifiable Component of Measurement Error (such as Operators or Machines) gives rise to measurements which differ by some fixed amount. Bias Effects are named after the particular Component of Measurement Error associated with the bias. That is, when different operators display detectably different average values for measurements of the same thing, an Operator Bias is said to exist. Whenever different machines display detectably different average values for measurements of the same thing, a Machine Bias is said to exist.

Bias Effects will usually be apparent on the Average Chart for the Basic EMP Study. In particular, an Operator Bias will show up as "shifts" in the running record which correspond to the different operators. If the running record shows shifts which correspond to the different machines in a study, then one would expect to find a detectable Machine Bias. Since such vertical displacements will usually be easy to see on the Average Chart for the Basic EMP Study, the following analysis is often simply a confirmatory procedure. It will determine if a suspected Bias Effect is likely to be real. Occasionally, however, a large product variation will make it hard to detect real Bias Effects on the Average Chart for the Basic EMP Study. Here the following analysis provides additional insight beyond that given by the Basic EMP Study.

The EMP Study for Bias Effects will use a Main-Effect Chart. This Main-Effect Chart uses the same data as the Basic EMP Study, but in a different manner. The steps for this EMP Study for Bias Effects are outlined and illustrated below.

The first step in building a Main-Effect Chart consists of obtaining some basic information from the regular control chart. The summary statistics and the estimate of the standard deviation of X will be among the information needed. (As noted below, when checking for a Bias Effect, this standard deviation will be that of Test-Retest Error.)

The second step in building a Main-Effect Chart is to rearrange the data into different subgroups and obtain the averages for these subgroups.

The third step in building a Main-Effect Chart is to estimate the standard deviation of the new subgroup averages and construct the appropriate control limits. The new subgroup averages are then plotted against these limits. If one or more of the new subgroup averages falls outside these control limits, then a detectable difference between the averages is said to exist. When the different averages correspond to the levels of some factor (a Main Effect), the presence of a detectable difference between the averages is said to represent the Main Effect for that factor. If none of the averages falls outside the control limits, then there is said to be no detectable Main Effect for that factor.

When a Main-Effect Chart is used there is no Range Chart.

To adapt the Main-Effect Chart to an EMP Study which will check for Bias Effects, proceed as follows:

Step I.   Obtain the following information from the Basic EMP Study:
    (a) Product Studied:
    (b) Measurement Studied:
    (c) Grand Average, $\bar{\bar{X}}$:
    (d) Average Range, $\bar{R}$:
    (e) Subgroup size, n:
    (f) Number of Subgroups, k:
    (g) Estimate of Standard Deviation for Test-Retest Erro, $\hat{\sigma}_e = \bar{R}/d_2$

Step II.   Identify the Potential Bias Effect To Be Considered:
    (a) Name of Component of Measurement Error:
    (b) Number of Levels for this Component, L:
    (c) Number of Observations per Level, N, (usually $N = nk/L$):
    (d) Names for Each Level of this Component:
    (e) Averages for Each Level of this Component:

Step III.   Compute Control Limits and Plot the Main-Effect Chart:
    (a) Find Test-Retest Error for the average of N measurements: $\hat{\sigma}_{e\,aver} = \dfrac{\hat{\sigma}_e}{\sqrt{N}}$

    (b) Control Limits are: $\bar{\bar{X}} \pm 3\,\hat{\sigma}_{e\,aver}$

    (c) Plot the Averages in Step II.(e) against the Limits in Step III.(b). If any of the Averages falls outside these control limits, then there is a detectable Bias Effect present in the data. Otherwise, there is no detectable Bias Effect which can be associated with the Component of Measurement Error being studied.

A worksheet for the EMP Study for Bias Effects is provided in the Appendix.

The use of a Main-Effect Chart is the simplest way to check for the presence of Bias Effects in the measurement process.

Separate analyses may be performed for each of the different Components of Measurement Error present in any given study.

Example 4.2:   EMP for Operator Bias: Gauge 109 Data:

 I. *From the Basic EMP Study for Gauge 109:*
  (a) Fitting Number 109
  (b) Gauge and Jig Number 109
  (c) Grand Average = 33.667
  (d) Average Range = 8.933
  (e) n = 2
  (f) k = 15
  (g) Estimated Test-Retest Error = $\hat{\sigma}_e$ = 7.92

 II. *Identify Potential Bias Effect:*
  (a) Operator Bias Effect
  (b) L = 3 levels
  (c) N = 10 observations/level
  (d) and (e)
  The average for Oper. A's ten measurements is 35.9.
  The average for Oper. B's ten measurements is 33.3.
  The average for Oper. C's ten measurements is 31.8.

 III. *Compute the Control Limits and Plot the Main-Effect Chart.*
  (a) $\hat{\sigma}_{e\,aver} = \dfrac{7.92}{\sqrt{10}} = 2.50$

  (b) *Control Limits are: 33.667 ± 7.50 = 26.17 to 41.17*

41.17 ·····························
         33.67
26.17 ·····························

Figure 4.2: EMP Study for Operator Bias: Gauge 109 Data:

There is no evidence of a detectable Operator Bias in these data. The differences between the Operator Averages can be accounted for in terms of Test-Retest Error alone.

Whenever a detectable Bias Effect is found, the next step is to estimate the bias for each level of the Component of Measurement Error. This is generally taken to be the difference between the average for each level and the Grand Average. (Of course, the sum of such estimates will always be zero.) If one can take steps to reduce or eliminate these biases, then the quality and consistency of the measurements will be improved. Failing this, one can at least compensate for the biases by using their estimated values.

EMP Studies for Inconsistency Effects

An Inconsistency Effect exists when the standard deviation of Test-Retest Error changes as the level of some other Component of Measurement Error changes. Inconsistency effects are named after the Component of Measurement Error that is associated with the inconsistency. That is, when different operators display detectably different standard deviations for Test-Retest Error the data are said to display an Operator Inconsistency. Obviously, this nomenclature is short for "an Inconsistency Effect that may be attributed to Operators."

In order to check for an Inconsistency Effect one must compare the dispersions found at different levels of some Component of Measurement Error. The mechanism for doing this will be a Chart for Mean Ranges.

Whenever an Inconsistency Effect exists it will usually be apparent on the Range Chart for the Basic EMP Study. In particular, an Operator Inconsistency will show up as "shifted" running records for the different operators, or as out-of-control range values for some operators. A simple way of checking for detectable differences between Mean Ranges for different Operators is to obtain a Chart for Mean Ranges. Thus, the EMP Study for Inconsistency Effects will consist of a Chart for Mean Ranges. This chart will also be based upon the data from the Basic EMP Study. The procedure for obtaining this Chart for Mean Ranges follows.

First, certain information must be obtained from the Basic EMP Study. This information includes the value for the (overall) Average Range, the number of subgroups, and the subgroup size. In addition, based upon this information, one must estimate the standard deviation of the subgroup ranges for the Basic EMP Study. This is done by using the formula:

$$\text{Est. SD(R)} = \frac{d_3 \bar{\bar{R}}}{d_2}$$

where d_2 and d_3 are based upon the subgroup size for the Basic EMP Study, and $\bar{\bar{R}}$ is the Overall Average Range from the Basic EMP Study. (The double-bar symbol is used here to help distinguish between the Overall Average Range found in the Basic EMP Study from the various Average Ranges which will be computed in the next part of this procedure.)

Second, the potential inconsistency effect is identified. The Component of Measurement Error of interest and the number of levels for this factor are specified. As before, let L denote the number of levels. Next, one will have to obtain an Average Range for each level of this factor. Let M denote the number of ranges used to find each of these Average Ranges, (usually M = k/L , although M may vary from level to level).

Third, the control limits are computed, and the Chart for Mean Ranges is plotted. If any of the Average Ranges fall outside the limits on this chart, there is a detectable difference between the Test-Retest Error for the different levels of the factor considered. This is an Inconsistency Effect. If none of the Average Ranges fall outside the limits, then there is no evidence of a systematic shift in the Test-Retest Error from one level of the factor to another, and no Inconsistency Effect is detected.

These steps are outlined in greater detail below:

Step I. Information from the Basic EMP Study
  (a) Product Studied
  (b) Measurement Studied
  (c) Overall Average Range $= \bar{\bar{R}}$
  (d) Subgroup Size $= n$
  (e) Number of Subgroups $= k$
  (f) Est. SD(R) $= \dfrac{d_3 \bar{\bar{R}}}{d_2}$

Step II. Potential Inconsistency Effect
  (a) Component of Measurement Error Considered
  (b) Number of Levels $= L$
  (c) Number of ranges for each level, M (usually $M = k/L$)
  (d) Names for each level
  (e) Average Ranges for each level

Step III. Compute Control Limits and Plot the Chart for Mean Ranges
  (a) Est. SD($\bar{R}$) $= \dfrac{\text{Est. SD(R)}}{\sqrt{M}}$
  (b) Control Limits are: $\bar{\bar{R}} \pm 3\,[\,\text{Est. SD}(\bar{R})\,]$
  (c) Plot the Average Ranges from Step II.(e) against the limits in Step III.(b).

When a detectable Inconsistency Effect is found, one should seek to identify the Assignable Cause for the different values of Test-Retest Error. If this cause cannot be found and eliminated, the measurement process must be considered to be different for each different level of the factor studied. In terms of an Operator Inconsistency Effect, this means that each operator has a different measurement process! In terms of a Machine Inconsistency Effect, this means that each machine has its own Test-Retest Error: in other words, the measurements from each machine display a different amount of measurement error!

A worksheet for EMP Studies for Inconsistency Effects is given in the Appendix.

The Gauge 109 Data will be used to illustrate the use of a Chart for Mean Ranges to check for Inconsistency Effects in the context of an EMP Study.

<u>Example 4.3:</u>   <u>EMP for Operator Inconsistency: Gauge 109:</u>

Step I.   *Information from the Basic EMP Study*

     (a)   *Product Studied: Fitting Number 109*

     (b)   *Measurement Studied: Gauge and Jig Number 109*

     (c)   *Overall Average Range* $= \bar{\bar{R}} = 8.933$

     (d)   *Subgroup Size* $= n = 2$

     (e)   *Number of Subgroups* $= k = 15$

     (f) $\text{Est. SD(R)} = \dfrac{d_3 \bar{\bar{R}}}{d_2} = \dfrac{(0.853)\ 8.933}{1.128} = 6.755$

(Note that the 0.853 and 1.128 are the appropriate values for n = 2.)

Step II.   *Potential Inconsistency Effect*

     (a)   *Check for Operator Inconsistency*

     (b)   *Number of Levels* $= L = 3$

     (c)   *Number of ranges for each level,* $M = k/L = 15/3 = 5$

     (d) and (e)

The Average Range for Operator A's five subgroups is 11.4

The Average Range for Operator B's five subgroups is 8.2

The Average Range for Operator C's five subgroups is 7.2

Step III.   *Compute Control Limits and Plot the Chart for Mean Ranges*

     (a) $\text{Est. SD}(\bar{R}) = \dfrac{\text{Est. SD(R)}}{\sqrt{M}} = \dfrac{6.755}{\sqrt{5}} = 3.021$

     (b)   *Control Limits are:*

$\bar{\bar{R}} \pm 3\,[\,\text{Est. SD}(\bar{R})\,] = 8.933 \pm 3\,[3.021] = -0.13\ to\ 17.996.$

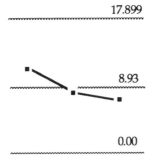

17.899

8.93

0.00

Figure 4.3: EMP Study for Operator Inconsistency: Gauge 109 Data

There is no evidence of a detectable Inconsistency Effect due to operators.   One must consider all three operators to have the same Test-Retest Error.

Since the estimate of the standard deviation of the Range Distribution requires both d_2 and d_3, these constants are given in the following table and in the Appendix.

Table 4.1
Constants Associated With the Range

n	d_2	d_3
2	1.128	0.853
3	1.693	0.888
4	2.059	0.880
5	2.326	0.864
6	2.534	0.848
7	2.704	0.833
8	2.847	0.820
9	2.970	0.808
10	3.078	0.797

The Chart for Mean Ranges may be used to check each of several different Components of Measurement Error for evidence of Inconsistency Effects.

Summary of EMP Studies

A complete EMP Study will begin with an Average and Range Chart in which the subgroups are arranged to include only Test-Retest Error within the subgroups. All other sources of variation, including other Components of Measurement Error, are made to appear between the subgroups. When it is appropriate, Main-Effect Charts may be constructed to check for Bias Effects, and Charts for Mean Ranges may be constructed to check for Inconsistency Effects. Generally, when Bias Effects or Inconsistency Effects are severe, they will be detectable on the Average Chart or the Range Chart for the Basic EMP Study. However, the Main-Effect Chart and the Chart for Mean Ranges provide a simple and easy way of checking for Bias Effects or Inconsistency Effects which might be overlooked on the Basic EMP Charts. Therefore, while these follow-up charts are optional, the best practice will be to include them whenever they make sense.

Five extended examples of EMP Studies are provided in Chapters 6 through 10.

The Use of ANOM With EMP Studies

The follow-up EMP Studies are focused on detecting and displaying those Bias Effects and Inconsistency Effects that may be present in the measurement process. Whenever evidence of any such effect is found, one must take action either to remove and eliminate the effect, or else to quantify and (if possible) compensate for the effect. Therefore, since action must follow discovery, one will be rather conservative in deciding that such an effect is present in the data. The Control Chart Approach outlined above is naturally conservative when applied to these follow-up EMP Studies.

The Analysis of Means (ANOM) approach may be used as an alternative for the Control Chart Approach. With the ANOM approach, the user must (1) determine the degrees of freedom for his estimate of SD(X), and (2) specify the overall α-level for the procedure. Having done this, the user can determine the appropriate H value to use in place of the customary control chart multiplier of 3.0. Thus, in return for some slight added complexity, the user can "fine-tune" the analysis to use a specified α-level.

In an exploratory analysis, this ability to choose a large α-level in order to increase the sensitivity of the analysis is a distinct advantage for ANOM. However, when the analysis must be confirmatory (i.e. conservative), the ANOM advantage over the Control Chart Approach quickly disappears. If one is truly conservative and chooses $\alpha = 0.01$, the value for H will often be in the neighborhood of 3.0. If one is less conservative and chooses $\alpha = 0.05$, one will, on the average, look for **non-existent** Bias Effects or Inconsistency Effects for about one analysis out of every 20. Given the nature of the actions involved, this false alarm rate will generally be unacceptable.

Thus, given the need for a conservative approach, the ANOM procedure will rarely produce results that differ from those obtained with the Control Chart Approach, yet it requires more work. Therefore, while the ANOM procedure can be properly used for the follow-up EMP Studies, it can hardly be recommended as advantageous. The same results can be obtained more easily with the Control Chart Approach.

Chapter Five

The Relative Usefulness of a Measurement

The standard deviation for Test-Retest Error provides an absolute measure of the basic Component of Measurement Error. However, it does not characterize the relative usefulness of that measurement for a specific product. Before such a relative measure of usefulness can be obtained, one will have to have an estimate of the product variation, and a framework for comparing Test-Retest Error with product variation. This framework, the basic relationships behind this framework, and a specific statistic for measuring the relative usefulness of a measurement, are given in this chapter.

To begin, assume that measurement error is independent of product variation. (Although the standard deviation for measurement error may change as the magnitude of the measurements changes, it is usually reasonable to assume that the measurement error standard deviation is constant within the range defined by the Natural Process Limits for a specified product, and that measurement errors are independent of product variation.) This independence of measurement error and product variation can be displayed in the following formula for the variance of a individual measurement:

Variance of Measurement = Variance of Product + Variance of Measurement Error

or, in symbols:

$$\sigma_m^2 = \sigma_p^2 + \sigma_e^2.$$

(This notation differs from that in the first edition.)

Since it is the variances which are additive, one can represent the relationship above by a right triangle. The standard deviation of a single measurement is equal to the length of the hypotenuse, while the product standard deviation and the measurement error standard deviation are equal to the lengths of the two legs.

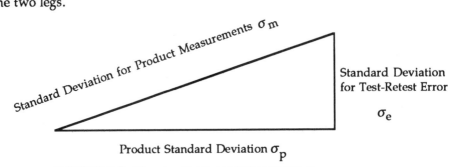

Figure 5.1: The Relationship Between Standard Deviations

(Since it is desirable to remove all identifiable sources of measurement variation such as Operator Effects, Machine Effects, or Lab Effects, from the measurement process, one may use the standard deviation for Test-Retest Error as the measurement error standard deviation.)

A unit change in either the Product Standard Deviation, or the standard deviation for Test-Retest Error, will not result in a unit change in the standard deviation for the Product Measurements.

As long as the product standard deviation is greater than the Test-Retest Error standard deviation, it will have the greater impact upon the standard deviation for product measurements. However, as one practices Continual Improvement, the product standard deviation will be reduced. As this happens, one will come to the point where the Test-Retest Error standard deviation dominates the standard deviation for product measurements. As this occurs it will become increasingly difficult to detect any further improvements in the production process, and one will need to work on improving the measurement process.

For this reason, it is helpful to have some measure of the relative usefulness of a given measurement for a specific product.

The traditional statistic for relative usefulness is the Intraclass Correlation Coefficient. This coefficient may be defined as the ratio of two variances:

$$\text{Intraclass Correlation Coefficient} = \rho_I = \frac{\text{Var.(Product)}}{\text{Var.(Measurement)}} = \frac{\sigma_p^2}{\sigma_m^2}$$

Inspection of this ratio suggests that one could interpret the Intraclass Correlation to be that proportion of the variation in the Product Measurements which is due to Product Variation.

Of course one can estimate this coefficient by using estimates for the two variances. When this is done, one has a Sample Intraclass Correlation Coefficient, r_I:

$$r_I = \frac{\text{Est. Var.(Product)}}{\text{Est. Var.(Measurement)}} = \frac{\hat{\sigma}_p^2}{\hat{\sigma}_m^2} = \frac{\hat{\sigma}_m^2 - \hat{\sigma}_e^2}{\hat{\sigma}_m^2} = 1 - \frac{\hat{\sigma}_e^2}{\hat{\sigma}_m^2}.$$

This Sample Intraclass Correlation Coefficient, r_I, is theoretically bounded by 0 and 1. As Test-Retest Error gets smaller relative to measurement variation, this coefficient goes to 1. Conversely, as Test-Retest Error gets larger relative to measurement variation, this coefficient will drop toward zero. (If this formula happens to give a negative estimate, it is simply an indication that the product variation is so small that it has been overwhelmed by the uncertainty in the estimates of σ_m^2 and σ_e^2.) While the Sample Intraclass Correlation Coefficient may be defined in other ways, the definition above is used because of its relative simplicity.

Unfortunately, while the definitions of both ρ_I and r_I are fairly simple, the interpretation of each is not very intuitive. The metric scale presented by the statistic is non-linear, which complicates the interpretation and clouds the relationships involved. For this reason, it is useful to transform this traditional measure of association into a Discrimination Ratio:

$$\text{Discrimination Ratio} = D_\rho = \sqrt{\frac{1+\rho}{1-\rho}} = \sqrt{\frac{2\sigma_m^2}{\sigma_e^2} - 1}$$

This Discrimination Ratio may be estimated by substituting estimates of the two variances:

$$\text{Estimated Discrimination Ratio} = D_R = \sqrt{\frac{2\hat{\sigma}_m^2}{\hat{\sigma}_e^2} - 1}$$

where estimates of these two variances are found as outlined below.

This Discrimination Ratio will characterize the relative usefulness of a given measurement for a specific product. Therefore, it augments the information provided by the standard deviation of Test-Retest Error. While the standard deviation of Test-Retest Error is an absolute measure of measurement uncertainty which applies within the range studied, the Discrimination Ratio is a relative measure of usefulness for a specific situation.

(The Discrimination Ratio defined above is slightly different from the Classification Ratio defined in the first edition of this book. This refinement will change only the smallest values, and both ratios will have the same interpretation.)

The estimate of σ_e^2 is usually obtained by squaring the standard deviation for Test-Retest Error. This estimate was discussed and illustrated in Chapters Two and Four (see pp. 14, 38 and 47).

The estimate of σ_m^2 will usually be obtained by squaring the value of $\bar{R}/d_2$ **obtained from a control chart for the specific product being considered.** This approach works because the quantity:

$$\text{Est. SD(X)} = \bar{R}/d_2$$

directly estimates the standard deviation of the individual measurements. Using such a control chart for this estimate will always be the preferred method for estimating σ_m^2.

An alternative technique for estimating σ_m^2 is available when the sample parts or batches for the Basic EMP Study have been selected in an objective manner from the product stream. Unfortunately, this technique yields an estimate for σ_m^2 having very few degrees of freedom. Consequently, both the estimate of σ_m^2 and the D_R statistic will be subject to considerable sampling variation. For this reason, this alternative technique should not be used when a product control chart is available.

This alternative technique will consist of directly estimating σ_m^2 from the sample parts or batches used in the Basic EMP Study. To do this, begin by finding the average for each sample part or batch in the Basic EMP Study. Thus, if there were several operators and/or several machines in the study, one would calculate the average for all measurements of the first sample part or batch by averaging across all the operators and all the machines used with the first sample. This would be done for each sample part or batch used in the study. Say that this results in p averages, each based upon N values: these p averages are then used to compute a sample variance, s_{aver}^2. Given this statistic, the estimate of σ_m^2 is given by the formula:

$$\hat{\sigma}_m^2 = s_{aver}^2 + \frac{N-1}{N} \hat{\sigma}_e^2$$

This estimate may then be used to compute an estimate of the Discrimination Ratio, D_R. This technique will be illustrated with the Gauge 109 data.

Another variation on this approach would consist of using $(R_{aver}/d_2)^2$ in place of the s^2 term in the formula above. The range would be found from the p sample averages just as the s^2 term was, and the value of d_2 would correspond to a "subgroup size" of p.

Both of these alternative techniques for estimating σ_m^2 use the data from the Basic EMP Study, rather than a control chart for the product being studied. Therefore, these two alternative techniques are built upon the assumption that the sample parts or batches used in the special study were objectively obtained from the product stream over an extended period of time. If the samples were obtained during a limited period of time, they may under-represent the product variation and deflate the Discrimination Ratio accordingly. If the samples were not objectively obtained from the product stream, but were instead prepared or selected because they covered some region of interest, then the formula above cannot be used, and any Discrimination Ratio based upon an estimate of σ_m^2 obtained from this formula will be worthless. Before the Discrimination Ratio can be meaningful it must incorporate a reasonable estimate of the product variation. As always, the best place to obtain such an estimate is the control chart for the product in question.

If an EMP Study is performed during the design phase of a manufacturing process it is likely that the parts or batches used will be prototypes. When this occurs, it will be quite difficult to estimate the variation of the production process. Therefore, the major purpose of an EMP Study at this phase should be the estimation of Test-Retest Error, rather than the estimation of a Discrimination Ratio.

The estimate of σ_e^2 *comes from the Range Chart for the Basic EMP Study:* $\bar{R}$ *was 8.933, and n = 2, so:*

$$\hat{\sigma}_e^2 \ = \ (\bar{R}/d_2)^2 \ = \ (8.933/1.128)^2 = 62.716.$$

The estimate of σ_m^2 *should properly come from a control chart for the Brass Fitting measurements.   When such a chart is not available, one can use the following approximation:*

$$\hat{\sigma}_m^2 \ = \ s_{aver}^2 + \frac{N\text{-}1}{N}\,\hat{\sigma}_e^2$$

where s_{aver}^2 *is the* s^2 *value obtained from the averages for each sample part or batch. For the Gauge 109 Data each of the five parts was measured six times, so N = 6. The five averages (one average for each part) are:*

$$35.833,\ 45.833,\ 12.167,\ 53.167,\ 21.333$$

so the formula above gives:

$$\hat{\sigma}_m^2 \ = \ 286.833 + (5/6)\,62.716 \ = \ 339.096.$$

Thus, the Discrimination Ratio for the Gauge 109 Data is estimated to be:

$$D_R \ = \ \sqrt{\frac{2\,\hat{\sigma}_m^2}{\hat{\sigma}_e^2} - 1} \ \ = \ \sqrt{\frac{2\,(339.096)}{62.716} - 1} \ \ = \ 3.13.$$

This value of 3.13 suggests that while the measurements made with Gauge 109 can detect some product variation, they cannot discriminate with any great precision. In fact, within the Natural Process Limits, the Gauge 109 measurements can divide the product into only three categories: high, medium and low. A justification of this interpretation of the Discrimination Ratio follows.

Interpreting the Discrimination Ratio

The following plots are shown here in order to provide some insight into just what the Discrimination Ratio is measuring. Since these plots use the same data that have already been plotted on the Average and Range Charts of the Basic EMP Study, they do not have to be drawn in practice unless the experimenter wishes to do so.

When a Basic EMP Study is performed, one could plot an Intraclass Correlation Plot instead of (or in addition to) the Average and Range Chart. The procedure for this special scatterplot follows.

When the Basic EMP Study has subgroups of size two, each subgroup will define two points in the XY plane. Let x_1 denote the first value in the subgroup, and let x_2 denote the second value in the subgroup, then plot the points (x_1, x_2) and (x_2, x_1) in the XY plane on regular graph paper. As this is done for each subgroup the result is a set of points which are symmetrically placed about the positive 45° line.

Consider the data from the EMP Study of Gauge 109 (p.38). The first subgroup had values of 34 and 45, while the second subgroup had values of 56 and 44. These two subgroups generate four points in the intraclass correlation plot as shown in Figure 5.2.

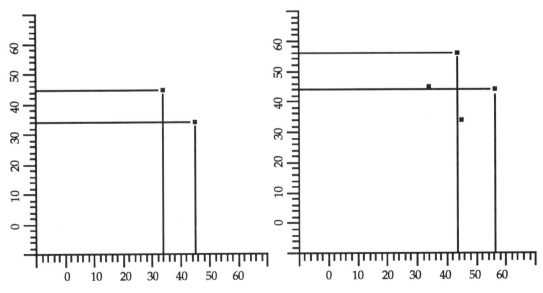

Figure 5.2: Plotting Points On The Intraclass Correlation Plot

Continuing in this manner, the plot will contain thirty points symmetrically distributed about the 45° line.

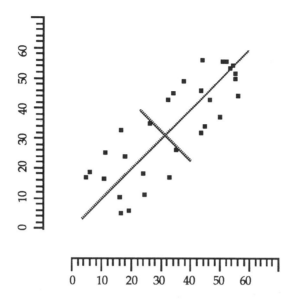

Figure 5.3: The Intraclass Correlation Plot for the Gauge 109 Data.

These points could be thought of as coming from some bivariate distribution. Assuming that it is a Bivariate Normal Distribution with equal variances and a correlation given by the Intraclass Correlation Coefficient, then the points on this plot could be thought of as fitting within some ellipse, and the ratio of the major axis of this ellipse to the minor axis would be given by the quantity

$$\sqrt{\frac{1+\rho}{1-\rho}}$$

The relationship between the ellipses and the correlation coefficient for the equivariant Bivariate Normal Distribution is shown in Figure 5.4.

The ellipses in Figure 5.4 provide theoretical models to use in interpreting the Intraclass Correlation Plots. The estimated Intraclass Correlation Coefficient for the Gauge 109 Data is 0.815.

If the measurements were perfect, then there would be no variation within the subgroups, all of the points would fall on the major axis, and the correlation would be 1.00. Departures from the major axis are strictly due to measurement error. At the same time, the elongation of the scatter diagram is due to the combination of product variation and measurement error. Thus, variation along the major axis is attributable to product variation and measurement error, while variation parallel to the minor axis is attributable to measurement error alone.

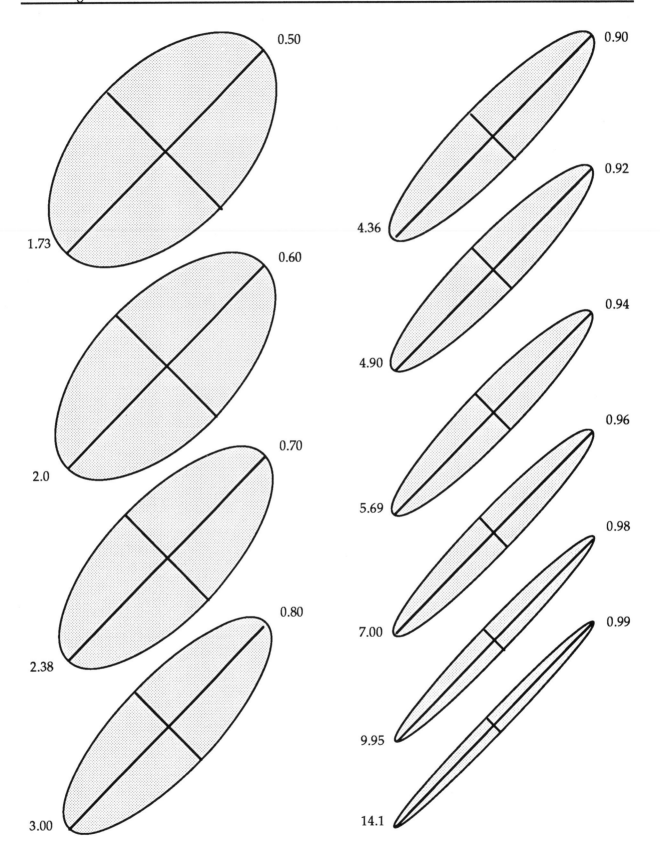

Figure 5.4: Discrimination Ratios, Ellipses and Correlations

A descriptive interpretation of the Discrimination Ratio can be obtained in the following manner. Since the minor axis defines the amount of uncertainty due to measurement error in one dimension, it will also do so in the other dimension. Therefore, allowing for measurement error, repeated measurements of the same thing will most likely fall somewhere within a square having a side equal to the minor axis. See, for example, the 6 points for Part 1 in Figure 5.5.

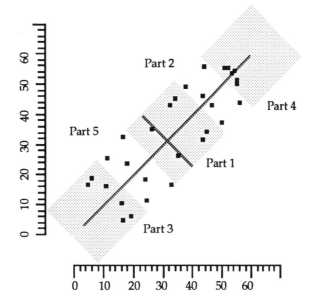

Figure 5.5: Clustering of Repeated Measurements

Squares such as these define regions within which variation is obscured by measurement error, making further discrimination difficult or impossible. If the ellipse was covered by a rectangle made up of nonoverlapping squares such as these, then the number of squares would define the number of product categories which could be established within the natural process limits using these measurements.

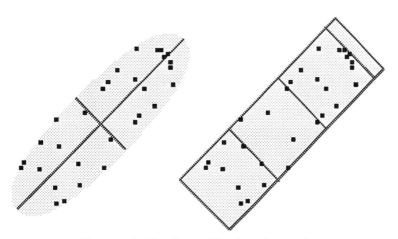

Figure 5.6: Number of Product Categories

Figure 5.6 shows the Intraclass Correlation Plot for the Gauge 109 data, the corresponding ellipse, and the squares which cover the ellipse. It takes slightly more than 3 squares to cover the ellipse, and the Discrimination Ratio for these data was estimated to be 3.13.

Thus, product variation exceeds measurement error to the same extent that the major axis exceeds the minor axis. **The ratio of the major axis to the minor axis defines the number of distinct product categories which could be established with the measurements while making allowance for measurement error, and this ratio is estimated by the Discrimination Ratio.** The greater this ratio, the more useful the measurement will be for that specific product. Likewise, the smaller this ratio, the more measurement error will dominate the measurement variation, and the less useful the measurements will be for that specific product.

It should be noted here that one measurement process might have considerably different Discrimination Ratios for different products.

The comparison between product variation and variation due to measurement error, which is quantified by the Discrimination Ratio, is also seen (approximately) on the Average Chart. The relationship between the Discrimination Ratio and the Average Chart in the Basic EMP Study is illustrated below.

The Average Chart for the Basic EMP Study for Gauge 109 is shown in Figure 5.7. The Discrimination Ratio for these data is 3.13.

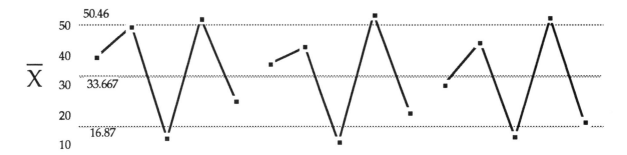

Figure 5.7: The Average Chart for the Basic EMP Study for Gauge 109

The Average Chart for another EMP Study (See Chapter 6 for the complete history) is shown in Figure 5.8. The Discrimination Ratio for these data is 10.2.

In general, the wider the band swept out by the running record becomes relative to the control limits, the greater the Discrimination Ratio. This relationship exists because the band swept out by the running record represents variation due to product variation and measurement error, while the width of the control limits represents that amount of variation which is obscured due to measurement error.

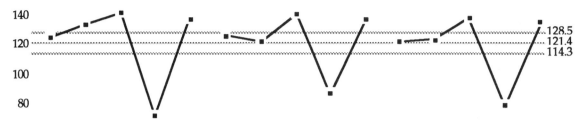

Figure 5.8: The Average Chart for the Second Gauge 130 EMP Study

The problem with a direct interpretation of these two dimensions on the Average Chart is that they represent the different quantities in different ways. Therefore, while these dimensions on the Average Chart do give a rough indication of the relative usefulness of the measurements for the product considered, the Discrimination Ratio provides a better measure of relative usefulness since it is based upon an equitable comparison.

Thus, since the Discrimination Ratio quantifies the relationship between Measurement Error and the Product Variation which is portrayed on the Average Chart in a Basic EMP Study, the Discrimination Ratio can be said to summarize and quantify the Average Chart itself.

A worksheet for computing Discrimination Ratios is provided in the Appendix.

Using Discrimination Ratios

Since the Discrimination Ratio quantifies the relative usefulness of a measurement process for a specific product it may be used to establish priorities with regard to improvement efforts. For example, given different processes, making different products, each with several product characteristics, should the resources available for process improvement be concentrated upon the production process or the measurement process? As long as the Discrimination Ratio indicates that the measurements can detect product variation it is best to concentrate upon the production process. When the Discrimination Ratio shows that a particular measurement cannot detect product variation it will be impossible to document whether the production process has improved, and it will be best to work on improving the measurement process. This latter course of action is certainly recommended when the Discrimination Ratio gets down in the region of 1.0 to 2.0. For simple measurements, it might be well to work on the measurement process when the Discrimination Ratio falls below 4.0 or so.

The Fallacy of Comparing Measurement Error to Specified Tolerance

The practice of comparing the standard deviation for Test-Retest Error to the Specified Tolerance can result in the incorrect allocation of resources. The following experience shows just how this can happen.

One day a statistician was called in to review an evaluation of a measurement system which had been performed according to the corporate manual. This study indicated that the product measurement had a standard deviation for Test-Retest Error of approximately 8 units ($\delta_e \approx 8.0$), while the Specified Tolerance for this product characteristic was only 20.0 units. Since the Test-Retest Error was 40% of the Specified Tolerance, the guidelines in the corporate manual suggested that the measurement process was unacceptable, and that it should be modified or replaced. The engineers could modify the measurement process, but the price tag of $1.6 million had made the plant manager pause. Thus, the corporate statistician was called in for a second opinion.

The statistician performed an EMP Study and found a Discrimination Ratio of 6.4. (The standard deviation for the product measurements was approximately 36 units ($\delta_m \approx 36.0$)). Therefore, even though Test-Retest Error was 40 percent as large as the Specified Tolerance, the current measurement process was quite capable of detecting product variation. Moreover, since the product variation was so great, most of the product was nonconforming. Spending $1.6 million to improve the measurement process would only help them to quantify more precisely how poorly they were doing! Clearly, rather than wasting time, resources, and effort on improving the measurement process, they needed to work on reducing the process variation!

The common practice of comparing the measurement variation with the Specified Tolerance does not consider the relative usefulness of the measurement in detecting product variation. While the situation may not always be so dramatic as the example above, the results will often be the same. Scarce manpower and resources, which would be better spent on process improvement, are misdirected to the improvement of the measurement process. The time to improve a measurement process is when it can no longer detect product variation. In short, when it has a small Discrimination Ratio.

This is why comparing Measurement Error with Specified Tolerance is inadequate. It does not fully characterize the measurement process, and it can lead to a misallocation of resources.

Chapter Six

The Data for Gauge 130

The EMP Study described in this chapter is actually composed of two different studies. In the first study there is evidence of inconsistency. As a consequence of removing this inconsistency and modifying the equipment, there was a definite improvement in the measurement process. This improvement was documented by the second study.

The Gauge 130 Measurement System consists of a jig designed to hold Fitting Number 130 and a measuring gauge. The quantity measured is the major dimension of the fitting, and these measurements are made in units of 0.001 mm. The values recorded are the amount by which the fitting exceeds the minimum specification.

The Basic EMP Study

This measurement system was evaluated by having each of three different individuals measure each of five different parts twice. The five parts were, in this instance, simply grabbed from the top of a basket of parts. The three individuals were Bill, the process engineer who designed the holding jigs, John, the operator who routinely used the gauge, and Terry, the quality control engineer conducting the study. The data and the Control Chart for the Basic EMP Study are given in Figure 6.1.

EMP Study of Gauge 130
Duplicate measurements of Brass Fitting Number 130

Oper.	Bill					John					Terry				
Part	1	2	3	4	5	1	2	3	4	5	1	2	3	4	5
Values	113	113	071	101	113	112	117	082	098	110	107	115	103	105	131
	114	106	073	097	130	112	107	083	099	108	109	122	086	109	090
$\bar{X}$	113.5	109.5	72.0	99.0	121.5	112.0	112.0	82.5	98.5	109.0	108	118.5	94.5	107	110.5
R	1	7	2	4	17	0	10	1	1	2	2	7	17	4	41

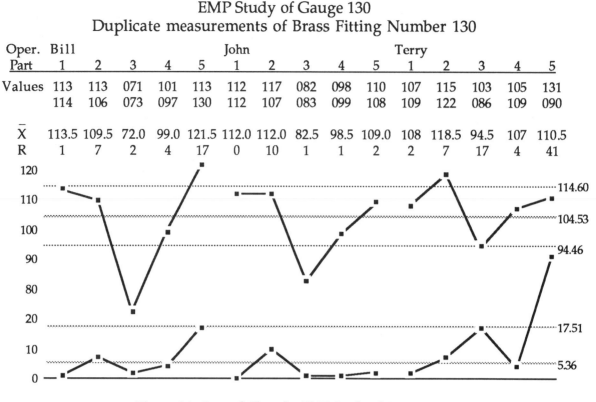

Figure 6.1: Control Chart for EMP Study of Gauge 130

Initially the Average Range was 7.733, giving an UCL_R of 25.3. One range value exceeded this upper limit. This out-of-control range value indicated a problem with the measurement process. Since it was Terry (the reliability engineer) who had the out-of-control range, he immediately went back to the floor and re-measured Part 5. As John (the operator) watched Terry measure Part 5, he observed that Terry was holding the part wrong. In particular, Terry's technique would allow the part to enter the jig incorrectly and thus skew the measurement. No one had told Terry about the special way to hold the part. Thus, the out-of-control range value had effectively and properly identified Terry to be an untrained operator!

Out-of-control ranges in a Basic EMP Study should always be investigated since they will almost always represent some inconsistency in the measurement process.

Having thus identified (and hopefully eliminated) the assignable cause for the out-of-control range, Terry deleted this range from further consideration. The revised value for the Average Range was 5.357. The Control Chart given in Figure 6.1 has limits based upon this Average Range.

The estimated standard deviation for Test-Retest Error is $\hat{\sigma}_e$ = 4.749 units, giving a Probable Error of ± 3.2 units. Thus, since the measurment unit is 0.001 millimeter, the effective resolution of this measurement procedure is approximately 3 thousandths of a millimeter.

In order to estimate the Discrimination Ratio for Gauge 130, begin with the standard deviation for Test-Retest Error. By squaring this estimated standard deviation one obtains an estimate of the Variance due to Measurement Error:

$$\hat{\sigma}_e^2 = (4.749)^2 = 22.553$$

Since a control chart for these dimensions on Fitting 130 was not available to the authors, the estimate of the Variance of the Measurements was obtained using the alternative method of estimating the variance of the averages for the five different parts. Based upon six measurements each, the averages for the five parts are:

$$111.167 \quad 113.333 \quad 83.0 \quad 101.5 \quad 113.667$$

The s_{aver}^2 value obtained from these five averages is 169.437. Therefore, the estimate of the variance of the measurements is:

$$\hat{\sigma}_m^2 = s_{aver}^2 + \frac{N-1}{N} \hat{\sigma}_e^2 = 169.437 + (5/6)(4.749)^2 = 188.23$$

Using this value one can compute an estimate of the Discrimination Ratio to be

$$D_R = \sqrt{\frac{2(188.23)}{22.553} - 1} = 3.96.$$

The five parts used in this study were a "grab sample" obtained from the top of a basket of fittings. Since the Discrimination Ratio computed above used the variation of these five parts to estimate $\hat{\sigma}_m^2$ one must interpret the Discrimination Ratio with care. If the five parts are excessively homogeneous, then the value of 3.96 may well be too small. On the other hand, if by some strange chance the five parts selected were abnormally different from each other, the value of 3.96 may be too large. (This latter situation is rather remote, but it could happen if, for instance, the five parts were produced on different heads and collected in the same basket in sequence.) It generally is best to sample the product stream rather than to draw samples from accumulations of product, and when there are parallel product streams, it is best to sample one stream at a time.

Because of the grab sample, the value of 3.96 is rather "soft." One does not know how good this ratio may be. However, it may be thought of as a lower bound on the discrimination of the measurements. Thus, it would appear that these measurements can identify at least four categories of product within the Natural Process Limits.

While the use of a grab sample does have an impact upon interpretation of the Discrimination Ratio, it does not affect the standard deviation of Test-Retest Error, $\hat{\sigma}_e$. This estimate and all the values obtained from it will be quite reliable.

Therefore, this Basic EMP Study has (1) identified an important aspect of performing these measurements: the proper holding technique, (2) given an absolute measure of the Test-Retest Error associated with this measurement process, (3) given a relative measure of the usefulness of these measurements for characterizing the product. Additionally, (4) aside from the out-of-control range, the Range Chart does not suggest any other measurement inconsistency, but (5) the Average Chart does suggest that there may be an Operator Bias Effect.

EMP Study for Operator Bias Effect

In order to check for possible Operator Bias Effects, a Main Effect Chart is used. The steps in the procedure are outlined below.

From the Basic EMP Study the Grand Average is 104.533 units, based on 15 subgroups of size n = 2. Using 14 subgroups the Average Range is 5.357 units, giving an estimate of the standard deviation for Test-Retest Error of 4.749 units.

To consider potential Operator Bias Effects one will need to obtain the average of all measurements for each operator separately. Here there are L = 3 operators, each having made 10 measurements.

> The average of Bill's ten values is 103.10.
> The average of John's ten values is 102.80.
> The average of Terry's ten values is 107.70.

Finally, control limits for the Main-Effect Chart are computed. The estimated Test-Retest Error for the average of ten measurements is:

$$\hat{\sigma}_{e\,aver} = \frac{\hat{\sigma}_e}{\sqrt{N}} = \frac{4.749}{\sqrt{10}} = 1.502.$$

Using this value, the control limits are:

$$\bar{\bar{X}} \pm 3\,\hat{\sigma}_{e\,aver} = 104.533 \pm 3\,(1.502) = 100.02 \text{ to } 109.03.$$

The Main-Effect Chart is shown in Figure 6.2.

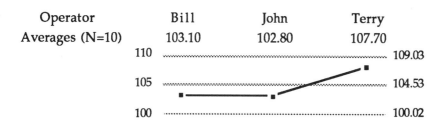

Figure 6.2: EMP Study for Operator Bias: Gauge 130

Figure 6.2 shows no evidence of a detectable Operator Bias. The differences between the Operator Averages can be accounted for in terms of Test-Retest Error alone.

EMP Study for Operator Inconsistency Effects

Although the Basic EMP Study did not suggest any Inconsistency Effects beyond the out-of-control range, the Chart for Mean Ranges will be obtained below. This procedure is included merely as an illustration of the technique. It would normally be used only when the Basic EMP Study suggested the possibility of some Inconsistency Effects.

From the Basic EMP Study, using k = 14 subgroups of size n = 2, the Overall Average Range is 5.357. Denote this value by $\bar{\bar{R}}$. Based upon this value, the estimated standard deviation of the Distribution of Subgroup Ranges is

$$\text{Est. SD(R)} = \frac{d_3 \bar{\bar{R}}}{d_2} = \frac{0.853 \,(5.357)}{1.128} = 4.051$$

To check for potential Operator Inconsistency Effects one will need to compute the Mean Range for the subgroups for each operator separately. Here there are L = 3 operators, each with 5 subgroups. However, since Terry's excessive range has been explained, only the first four of his subgroups will be used.

> The Average Range for Bill's five subgroups is 6.20.
> The Average Range for John's five subgroups is 2.80.
> And the Average Range for Terry's four subgroups is 7.50.

For the average of 5 ranges, the estimated standard deviation of $\bar{R}$ is:

$$\text{Est. SD}(\bar{R}) = \frac{\text{Est. SD(R)}}{\sqrt{M}} = \frac{4.051}{\sqrt{5}} = 1.812,$$

giving control limits for the Chart for Mean Ranges of:

$$\bar{\bar{R}} \pm 3\,[\,\text{Est. SD}(\bar{R})\,] = 5.357 \pm 3\,(1.812) = 0.0 \text{ to } 10.79.$$

Use these limits for Bill's and John's Average Ranges.

For the average of 4 ranges, the estimated standard deviation of $\bar{R}$ is:

$$\text{Est. SD}(\bar{R}) = \frac{\text{Est. SD(R)}}{\sqrt{M}} = \frac{4.051}{\sqrt{4}} = 2.026,$$

giving control limits for the Chart for Mean Ranges of:

$$\bar{\bar{R}} \pm 3\,[\,\text{Est. SD}(\bar{R})\,] = 5.357 \pm 3\,(2.026) = 0.0 \text{ to } 11.44.$$

Use these limits for Terry's Average Range. The Chart for Mean Ranges is shown in Figure 6.3.

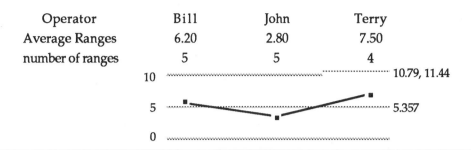

Figure 6.3: EMP Chart for Operator Inconsistency: Gauge 130

Since none of the operators has an Average Range which falls above the upper limit in Figure 6.3, there is no detectable evidence of Operator Inconsistency beyond that indicated by the out-of-control range noted on the original range chart.

Thus, this EMP Study of the Gauge 130 measurement process revealed the importance of holding the fittings properly, established the effective resolution of the measurements to be ± 3 thousandths of a millimeter, and suggested that the Discrimination Ratio is at least 4. In addition to all of these benefits, a side effect of the EMP Study was Terry's suggestion to Bill that they needed to change the load spring in the jig.

The load spring was therefore changed, and two days after the first EMP Study a second EMP Study was conducted. A fresh grab sample of five parts was obtained, and the same three operators participated in the new study. The data and the Control Charts are shown in Figure 6.4.

The effect of the change in the load spring is best seen in the Range Chart. The Average Range dropped to 3.80 units, giving an estimated standard deviation for Test-Retest Error of 3.369 units. Thus, changing the load spring resulted in a 29% reduction in Test-Retest Error. The Probable Error for this measurement process dropped to ± 0.0023 mm.

The Discrimination Ratio for these data is found in the same way as for the first study. Once again the fact that the five parts were a grab sample, rather than a sample selected from the product stream over an extended period of time, will tend to limit the utility of the Discrimination Ratio. However, the variation in the grab sample will at least give a rough idea of the product variation, so we can proceed to obtain a Discrimination Ratio.

EMP Study No. 2 of Gauge 130:
Duplicate measurements of a Brass Fitting

Oper.	Bill					John					Terry				
Part	1	2	3	4	5	1	2	3	4	5	1	2	3	4	5
Values	124	136	144	75	138	124	124	139	90	137	123	126	140	74	134
	125	131	140	70	137	128	121	142	85	138	121	120	136	85	136
$\bar{X}$	124.5	133.5	142	72.5	137.5	126	122.5	140.5	87.5	137.5	122	123	138	79.5	135
R	1	5	4	5	1	4	3	3	5	1	2	6	4	11	2

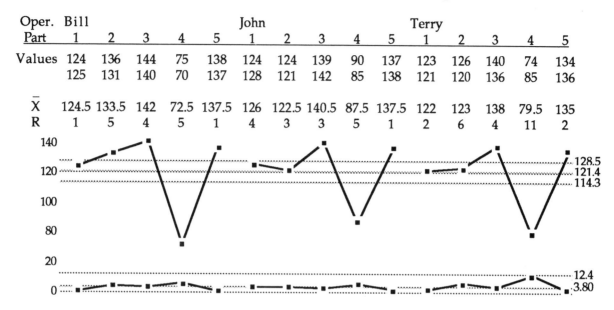

Figure 6.4: EMP Study No. 2 of Gauge 130

From the Second EMP Study shown in Figure 6.4:

$$\hat{\sigma}_e^2 = (\bar{R}/d_2)^2 = (3.369)^2 = 11.350$$

The estimate of $\hat{\sigma}_m^2$ is obtained from the averages for each part. The five parts were measured six times each, and their averages are:

$$124.167, \quad 126.333, \quad 140.167, \quad 79.833, \quad 136.667.$$

From these values we obtain:

$$s^2_{\text{aver}} = 586.271$$

so that the estimate of the standard deviation of the product measurements is:

$$\hat{\sigma}_m^2 = s^2_{\text{aver}} + \frac{N-1}{N} \hat{\sigma}_e^2 = 586.271 + (5/6)\,11.350 = 595.729.$$

Thus, the Discrimination Ratio for the Gauge 130, Study 2 Data is estimated to be:

$$D_R = \sqrt{\frac{2\,(595.729)}{11.350} - 1} = 10.2.$$

This value of 10.2 suggests that the measurements made with Gauge 130 are quite satisfactory for measuring these brass fittings. These measurements appear to be capable of distinguishing between fittings to the extent that one could establish ten distinct categories for the fittings within the Natural Process Limits.

It should be observed that while a portion of the increase in the Discrimination Ratio came from the reduced Test-Retest Error, the remainder of the increase was due to the increased variation between the five parts in the grab sample. This is why it is much better to estimate the $\hat{\sigma}_m^2$ value from a control chart for the product. Estimates obtained from such control charts will inevitably be more stable than any estimate based upon the sample parts or batches in the EMP Study.

Since there was no suggestion of Operator Bias or Operator Inconsistency in the Second Basic EMP Study, no follow-up EMP Studies were done.

Chapter Seven

Two Methods for Measuring Viscosity

The EMP Study described in this chapter compares two different measuring methods using studies with different subgroup sizes. Emphasis is given to the interpretation of the descriptive measures as a basis for making comparisons.

The viscosity of a given fluid may be measured in different ways. Two of the more common techniques are the U-Tube Method and the Cone and Plate Technique. A laboratory in one chemical company performed the following EMP Studies in order to assess the Test-Retest Error for these two different techniques as they were currently being used to measure the viscosity of Product 20F

Sample batches of Product 20F were obtained from four different lots. Each of these batches was large enough to allow for multiple viscosity tests to be performed. Prior to the selection of a subsample from any one of these sample batches the complete batch was stirred and agitated in order to homogenize the batch to the greatest extent possible.

Only one operator participated in these EMP Studies, and each study was performed on a single test stand, so there is only lot-to-lot variation and Test-Retest Error present in these data. Both EMP Studies are shown in Figure 7.1. The data are listed in units of 1000 centistokes.

Product 20F Measurement Error Study

Method	U-Tube				Cone and Plate			
Lot No.	43	51	59	52	43	51	59	52
	28.97	29.85	29.67	29.36	29.3	30.3	29.8	30.3
	28.73	30.19	29.73	29.94	29.2	30.3	29.9	30.3
					29.3	30.3	29.8	30.2
					29.2	30.2	29.7	30.1
					29.2	30.2	29.9	30.1
$\bar{X}$	28.85	30.02	29.70	29.65	29.24	30.26	29.82	30.20
R	0.24	0.34	0.06	0.58	0.1	0.1	0.2	0.2

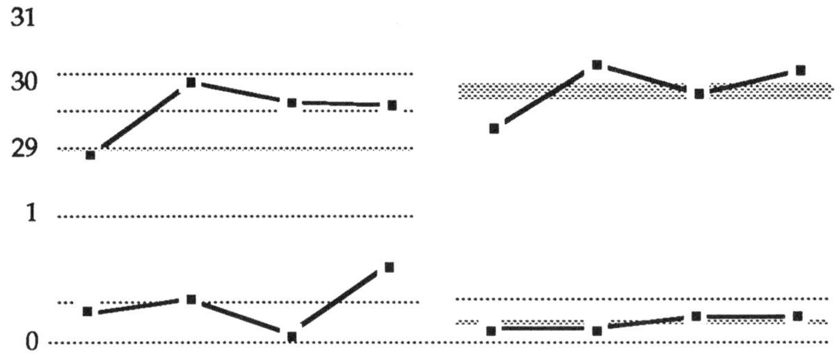

Figure 7.1: EMP Studies for Viscosity Measurements for Product 20F
(data listed in units of 1000 centistokes)

Since the limits for the Average Chart for subgroups of size n = 5 are naturally tighter than the limits for subgroups of size n = 2, it is difficult to make direct comparisons between these charts. Therefore, the computed summary values must be used to compare the two measuring techniques.

For the U-Tube data, using k = 4 subgroups of size n = 2, the Grand Average is 29.555, and the Average Range is 0.305. This gives an Upper Control Limit for Averages of 30.128, a Lower Control Limit for Averages of 28.982, and an Upper Control Limit for Ranges of 0.997.

The estimated standard deviation for Test-Retest Error for the U-tube method is $\bar{R}/d_2 =$ 0.305/1.128 = 0.270 units, or 270 centistokes. This makes the Probable Error ± 0.181 units (181 centistokes). Since the effective resolution of these measurements is ± 181 centistokes, it is excessive to record the viscosity to the nearest 10 centistokes (the present unit of measurement).

The Discrimination Ratio for the U-Tube measurements is D_R = 2.6. This value was found as follows: From the Test-Retest Error:

$$\hat{\sigma}_e^2 = (0.270)^2 = 0.072900.$$

From the Subgroup (Lot) Averages:

$$s^2_{aver} = 0.247767,$$

so that:

$$\hat{\sigma}^2_m = s^2_{aver} + \frac{N-1}{N}\hat{\sigma}^2_e = 0.247767 + (1/2)\,0.072900 = 0.284217.$$

Using these values, the Discrimination Ratio is found to be:

$$D_R = \sqrt{\frac{2\,(.284217)}{0.072900} - 1} = 2.607.$$

Thus, it would appear that the U-Tube measurements can detect about three categories of Product 20F within the Natural Process Limits. (Of course this estimate is only as good as the representation of product variation provided by the four lots in this study. A much better approach would be to estimate $\hat{\sigma}^2_m$ from a control chart for U-Tube measurements of Product 20F.)

For the Cone and Plate data, using k = 4 subgroups of size n = 5, the Grand Average is 29.880, and the Average Range is 0.15. This gives an Upper Control Limit for Averages of 29.967, a Lower Control Limit for Averages of 29.793, and an Upper Control Limit for Ranges of 0.317.

The estimated standard deviation for Test-Retest Error for the Cone and Plate method is $\bar{R}/d_2$ = 0.15/2.326 = 0.064 units, or 64 centistokes. This makes the Probable Error ± 0.043 units (43 centistokes). However, these data were recorded to the nearest 100 centistokes. A quick check of the Range Chart shows only four possible values for the range within the control limits (0.0, 0.1, 0.2, and 0.3). Since the subgroup size is n = 5, this is an indication of Inadequate Measurement Units (see p. 8). Thus, the round-off error exceeds the estimated Probable Error, and the effective resolution is the same as the measurement unit. The Cone and Plate measurements are good to the nearest 100 centistokes (and they have the potential to be even better). Clearly, the Cone and Plate measurements need to be recorded to at least one additional digit.

The Discrimination Ratio for the Cone and Plate measurements is D_R = 10.4. This value is found as follows: From the Test-Retest Error:

$$\hat{\sigma}^2_e = (.064)^2 = 0.004096.$$

From the Subgroup (Lot) Averages:

$$s^2_{aver} = 0.220000,$$

so that:

$$\hat{\sigma}^2_m = s^2_{aver} + \frac{N-1}{N}\hat{\sigma}^2_e = 0.220000 + (4/5)\,0.004096 = 0.223277.$$

Using these values, the Discrimination Ratio is found to be:

$$D_R = \sqrt{\frac{2\,\hat{\sigma}_m^2}{\hat{\sigma}_e^2} - 1} = \sqrt{\frac{2\,(.223277)}{0.004096} - 1} = 10.393.$$

This Discrimination Ratio for the Cone and Plate Data may be inflated due to the Inadequate Measurement Units. One of the effects of Inadequate Measurement Units is an artificial reduction in the Average Range. This will deflate the estimate of the standard deviation of Test-Retest Error, and inflate the Discrimination Ratio. A lower limit on the Discrimination Ratio may be obtained by setting the standard deviation for Test-Retest Error equal to the measurement unit (100 centistokes) and re-computing. The value obtained in this manner is, in this case:

$$\hat{\sigma}_m^2 = s_{aver}^2 + \frac{N-1}{N}\,\hat{\sigma}_e^2 = 0.220000 + (4/5)\,(0.1)^2 = 0.228000$$

and

$$\text{Minimum } D_R = \sqrt{\frac{2\,(.228)}{0.01} - 1} = 6.7.$$

Therefore, the Cone and Plate Method can discriminate at least 7 product categories in the same region where the U-Tube Method can discriminate at most 3 categories. Since both of these estimates are based upon the same lots, the relative utility of the two techniques is correctly shown regardless of the amount of variation in the product stream.

In all, these studies suggest that when it comes to measuring Product 20F at viscosities in the neighborhood of 30,000 centistokes, the Cone and Plate Method is much better than the U-Tube Method. This study does not attempt to make any statements regarding the usefulness of these two measurement methods under any other operating conditions.

Since these EMP Studies did not include any Component of Measurement Error besides Test-Retest Error, there is no need to perform analyses for Bias Effects or Inconsistency Effects.

Chapter Eight

The Truck Spoke Data

The EMP Study described in this chapter found Operator Inconsistency, Operator Bias, and Inadequate Measurement Units while still establishing that the measurement was useful for characterizing the product.

The insert for the spoke of a truck steering wheel is measured by using a jig and several measuring gauges. Each insert that is selected for measurement has nine different dimensions recorded. The following data are for Dimension No. 2.

Four operators routinely used this jig, so all four were asked to participate in these EMP Studies. Five parts were used. These five parts were selected on each of five different days. This was done in an effort to obtain a cross-section of product variation in the studies.

The dimensions were measured to the nearest half-thousandth of an inch (0.0005 in.). The values shown in Figure 8.1 were coded to make them easier to use. The gauge for Dimension No. 2 was set up so that a reading of 0 was equal to 1.5400 inch. The data were then recorded in units of 0.0001 inch in order to avoid having decimal values in the raw data. Thus, a recorded value of 45 represents a measured dimension of 1.5445 inch. Each Operator measured each part twice, giving k = 20 subgroups of size n = 2.

EMP Study for Truck Spoke Dimension No. 2

Oper	A					B					C					D				
Part	1	2	3	4	5	1	2	3	4	5	1	2	3	4	5	1	2	3	4	5
Values	20	20	25	50	45	20	15	15	45	35	20	20	25	45	40	0	10	10	30	20
	15	25	25	50	40	20	10	10	20	40	15	20	25	50	40	5	5	10	25	35
$\bar{X}$	17.5	22.5	25	50	42.5	20	12.5	12.5	32.5	37.5	17.5	20	25	47.5	40	2.5	7.5	10	27.5	27.5
R	5	5	0	0	5	0	5	5	25	5	5	0	0	5	0	5	5	0	5	15

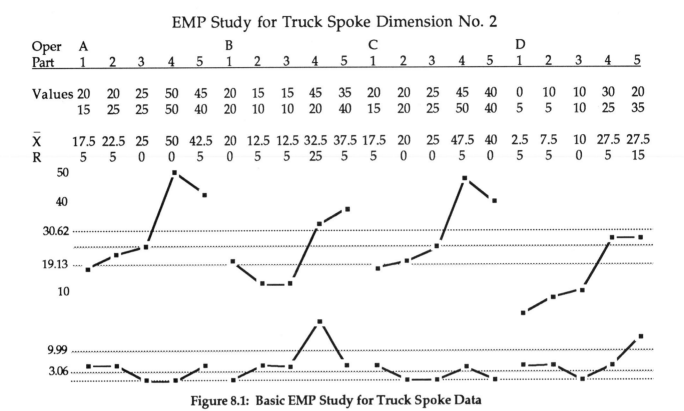

Figure 8.1: Basic EMP Study for Truck Spoke Data

Initially, the Average Range is 4.75 and the Upper Control Limit for Ranges is 15.52. Subgroup Nine has a range of 25 units. This out-of-control range is definite evidence of inconsistency in the measurement process.

Deleting this out-of-control range reduces the Average Range to 3.684, giving an Upper Control Limit of 12.04. Now the range for Subgroup 20 falls above the control limit.

Deleting the range for subgroup 20 gives an Average Range of 3.056 units and an Upper Control Limit of 9.986. No further points appear to be out of control on the Range Chart. These two out-of-control ranges suggest the possibility of an Operator Inconsistency Effect in these measurements.

However, while the initial control limits were probably inflated by the out-of-control ranges, the revised control limits reveal a different problem: there are only two possible values for the ranges within the control limits, namely 0 and 5. Thus, following the deletion of the out-of-control ranges, the remaining variation within the subgroups is found to be too small to be reliably detected by measurement units used.

So while the initial Average Range of 4.75 probably erred on the large side, the final Average Range of 3.056 probably errs on the small side. Estimates of the standard deviation for Test-Retest Error based on these two values are, respectively, 4.2 units and 2.7 units. The first of these is probably too large, and the last is probably too small.

Since the data display Inadequate Measurement Units, it is safe to say that the Test-Retest Error is definitely smaller than the measurement unit. Thus, while the effective resolution of these measurements is 0.0005 in., the measurement process is capable of even finer resolution. In order to realize this smaller resolution, smaller measurement units would have to be used.

While Inadequate Measurement Units will undermine a regular control chart for a production process, it is not so detrimental in an EMP Study. First of all, with an EMP Study we expect to find the Average Chart to be "out-of-control." Additional out-of-control points will not greatly change the interpretation of the Average Chart. Next, with an EMP Study, an indication of Inadequate Measurement Units places an upper bound on the standard deviation for Test-Retest Error, while it specifies that the actual resolution of the measurements is equal to the stated measurement unit. Since all of these side effects provide information about the measurement process, the presence of Inadequate Measurement Units does not undermine an EMP Study. Therefore, the final value for the Average Range was used to obtain control limits for the Average Chart shown in Figure 8.1.

The Average Chart suggests that these measurements are quite capable of detecting product variation. Moreover, the Average Chart shows a definite Operator Bias. The values obtained by Operators A and C are larger than those obtained by Operators B and D for the same parts.

Since Operators B and D also have occasional out-of-control ranges, it is likely that both Operator Bias Effects and Operator Inconsistency Effects are present in these data. In order to examine the data for these Operator Effects two follow-up EMP Studies will be performed.

EMP Study for Operator Bias Effect

From the Basic EMP Study using k = 20 subgroups of size n = 2, the Grand Average is 24.875. Based on 18 subgroups the Average Range is 3.056, and the estimate of the standard deviation of Test-Retest Error is 2.7 units.

The four operators in the study each made ten measurements. The averages for each operator are:

Operator A averaged 31.5 units,
Operator B averaged 23.0 units,
Operator C averaged 30.0 units, and
Operator D averaged 15.0 units.

Given that each of these averages is based upon ten measurements, the estimated Test-Retest Error for these averages is:

$$\hat{\sigma}_{e\,aver} = \frac{\hat{\sigma}_e}{\sqrt{N}} = \frac{2.7}{\sqrt{10}} = 0.85 \text{ units.}$$

Therefore, the control limits for a Main Effect Chart will be:

$$\bar{\bar{X}} \pm 3\,\hat{\sigma}_{e\,aver} = 24.875 \pm 3\,(0.85) = 22.31 \text{ to } 27.44.$$

Operator	A	B	C	D
Average	31.5	23.0	30.0	15.0

Figure 8.2: EMP for Operator Bias: Truck Spoke Data

Figure 8.2 shows three of the Operator Averages outside the limits, thus there is detectable Operator Bias present in these data. We therefore estimate the size of these biases:

Operator A's bias is 6.6 ten-thousandths above the grand average (31.5 – 24.875 = 6.625).

Operator B's bias is 1.9 ten-thousandths below the grand average (23.0 – 24.875 = –1.875).

Operator C's bias is 5.1 ten-thousandths above the grand average (30.0 – 24.875 = 5.125).

Operator D's bias is 9.9 ten-thousandths below the grand average (15.0 – 24.875 = –9.875).

If one cannot remove these biases, one can at least quantify them and possibly adjust for them. The source of these bias effects will be described below. Since they were able to remove these biases, it was unnecessary to adjust for them.

EMP Study for Operator Inconsistency Effects

The out-of-control ranges are definite indicators of measurement inconsistency. Therefore, there may be little need for this follow-up study. However, if one eliminates the two out-of-control ranges, this study may be used to address the question of whether there is any further inconsistency in the measurement process beyond that indicated by the excessive ranges.

From the Basic EMP Study using only 18 of the 20 subgroups of size n = 2, the Overall Average Range is:

$$\bar{\bar{R}} = 3.056 \text{ units}$$

which gives an estimate of the standard deviation of the Distribution of Subgroup Ranges of:

$$\text{Est. SD(R)} = \frac{d_3 \bar{\bar{R}}}{d_2} = \frac{0.853\,(3.056)}{1.128} = 2.31 \text{ units.}$$

The average ranges are:

The Average Range for Operator A's five subgroups is 3.0.

The Average Range for Operator B's four subgroups is 3.75.

The Average Range for Operator C's five subgroups is 2.0.

The Average Range for Operator D's four subgroups is 3.75.

The control limits for the Chart for Mean Ranges are found as follows:

For Operators A and C:

$$\text{Est. SD}(\bar{R}) = \frac{\text{Est. SD}(R)}{\sqrt{M}} = \frac{2.31}{\sqrt{5}} = 1.03 \text{ units}$$

so that:

$$\bar{\bar{R}} \pm 3\,[\,\text{Est. SD}(\bar{R})\,] = 3.06 \pm 3\,(1.03) = 3.06 \pm 3.09 = 0.0 \text{ to } 6.15 \text{ units}.$$

For Operators B and D:

$$\text{Est. SD}(\bar{R}) = \frac{\text{Est. SD}(R)}{\sqrt{M}} = \frac{2.31}{\sqrt{4}} = 1.16 \text{ units}$$

so that:

$$\bar{\bar{R}} \pm 3\,[\,\text{Est. SD}(\bar{R})\,] = 3.06 \pm 3\,(1.16) = 3.06 \pm 3.48 = 0.0 \text{ to } 6.54 \text{ units}.$$

These limits are shown on the Chart for Mean Ranges given in Figure 8.3.

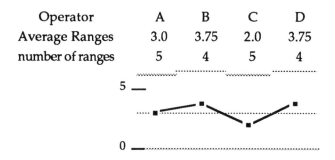

Operator	A	B	C	D
Average Ranges	3.0	3.75	2.0	3.75
number of ranges	5	4	5	4

Figure 8.3: EMP for Operator Inconsistency: Truck Spoke Data

Figure 8.3 does not show a detectable difference between the Average Ranges. However, the original control chart for ranges shows two out-of-control ranges. Thus, Operators B and D will have occasional inconsistencies, but aside from these occasional problems, there is no evidence of a systematic difference in Test-Retest Error due to different operators. As long as Operators B and D can avoid occasional bad measurements, they will have about the same amount of Test-Retest variation as Operators A and C.

Given the definite Operator Effects present in these data it was natural to check on the techniques used by the different operators. Operators B and D had one technique, and Operators A and C used another. In particular, Operators A and C pushed on the parts to "seat" them in the measuring jig, while Operators B and D pulled on the parts to seat them in the jig. Since the jig was designed so that the parts are removed by pulling on them, the technique of Operators B and D was incorrect. By pulling on the part, they were occasionally "unseating" the part rather than seating it! If the part was not properly seated, then a bad measurement would result, with the accompanying large range. Moreover, by pulling on the parts, Operators B and D were changing all of the measurements by an amount equal to the movement of the part in the jig. Thus, both the Operator Bias Effect and the Operator Inconsistency Effect can be explained in terms of the difference in techniques. With the help of the control charts from the EMP Study it was possible to get the Operators B and D to start using the technique of Operators A and C.

Finally, using the five parts in this EMP Study, the Discrimination Ratio is found as follows: assuming the best case, the smallest estimate of the standard deviation for Test-Retest Error gives:

$$\hat{\sigma}_e^2 = 7.29.$$

The averages for the five parts used are:

$$14.375, \quad 15.625, \quad 18.125, \quad 39.375, \quad 36.875.$$

Each of these averages is based upon eight measurements. Using these five averages we find:

$$s_{aver}^2 = 148.906$$

giving:

$$\hat{\sigma}_m^2 = s_{aver}^2 + \frac{N-1}{N} \hat{\sigma}_e^2 = 148.906 + (7/8) \, 7.29 = 155.285.$$

From these values we find:

$$D_R = \sqrt{\frac{2 \, (155.285)}{7.29} - 1} = 6.45.$$

Thus, if these five parts properly reflect the product variation, then at best these measurements can distinguish about 6 or 7 categories of product within the Natural Process Limits.

Since the estimate of $\hat{\sigma}_e^2$ used above is probably on the small side, the Discrimination Ratio is likely to be a little too large. An idea of just how much too large may be obtained by calculating a worst-case Discrimination Ratio. This worst-case ratio will be found using the larger estimate of Test-Retest Error.

The larger value for the standard deviation of Test-Retest Error was 4.2 units. Squaring this value and combining it with the values above gives a Discrimination Ratio of 4.20. Thus, it would appear that this measurement system is capable of establishing at least 4, and possibly 7 categories for the Truck Spoke Dimension No. 2 measurements.

Chapter Nine

The Data for Polymer 62S

The EMP Study described in this chapter checks for both Method Effects and Operator Effects. It detects a definite bias due to the measurement method employed.

The Gum Activity of Polymer 62S may be measured using three different techniques. The technique which is currently used with production data is Method G. In addition to Method G there is also Method H and Method RP. These three measurement methods were compared by means of an EMP Study.

In order to keep the EMP Study reasonably small only two sample batches of Polymer 62S were used. Both of these sample batches were especially prepared for this study, with one sample representing the high end of the range of Gum Activity, and the other sample representing the low end of the range of Gum Activity.

Three operators participated in this study, with each operator using all three measurement methods on each of the two samples. Each combination of operator, method, and sample occurred twice during the course of the study, giving 36 measurements arranged in k = 18 subgroups of size n = 2.

EMP Study for Gum Activity for Polymer 62S

Method	H						G						RP					
Operator	A		B		C		A		B		C		A		B		C	
Sample	1	2	1	2	1	2	1	2	1	2	1	2	1	2	1	2	1	2
	13.60	34.37	13.02	28.14	11.87	34.37	7.93	27.41	13.36	33.24	8.17	31.25	10.7	29.2	10.0	29.1	10.1	31.9
	11.36	31.69	13.46	31.69	14.76	35.72	8.30	30.45	12.71	29.42	10.13	29.80	8.6	31.1	9.1	31.1	7.8	29.7
Averages	12.48	33.03	13.24	29.92	13.32	35.04	8.12	28.93	12.54	31.33	9.15	30.53	9.65	31.15	9.55	30.10	8.95	30.80
Ranges	2.24	2.68	0.44	3.55	2.89	1.35	0.37	3.04	0.35	3.82	1.96	1.45	2.1	1.9	0.9	2.0	2.3	2.2

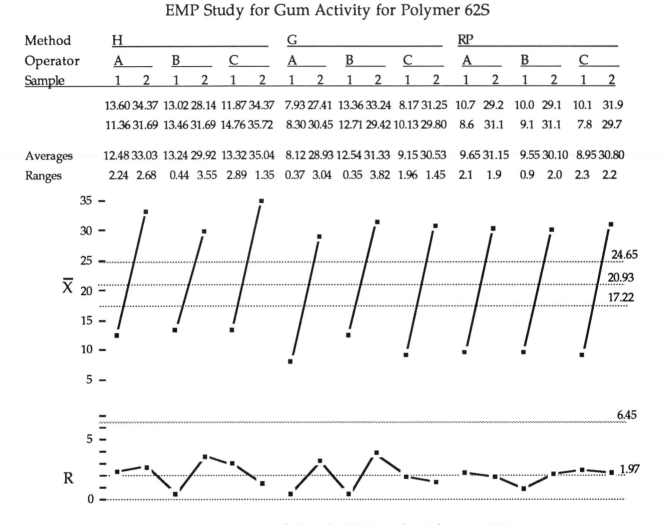

Figure 9.1: Control Chart for EMP Study: Polymer 62S Data

The Grand Average is 20.934, and the Average Range is 1.974 units. Since the Range Chart does not display any out-of-control points, it is reasonable to assume that the Test-Retest Error is essentially the same across all three measurement methods. Based on the Average Range, an estimate of the standard deviation for Test-Retest Error is 1.75 units, giving a Probable Error of ± 1.17 units.

However, since a comparison between the three measurement methods is one of the purposes of this study, it might be of interest to specifically examine the Chart for Mean Ranges before we conclude that the three methods have the same Test-Retest Error.

EMP Study for Method Inconsistency Effects

From the Basic EMP Study using k = 18 subgroups of size n = 2, the Overall Average Range is:

$$\bar{\bar{R}} = 1.9744 \text{ units.}$$

which gives an estimate of the standard deviation of the Distribution of Subgroup Ranges of:

$$\text{Est. SD(R)} = \frac{d_3 \bar{\bar{R}}}{d_2} = \frac{0.853\,(1.9744)}{1.128} = 1.493 \text{ units.}$$

The average ranges are:

The Average Range for Method H's six subgroups is 2.192.

The Average Range for Method G's six subgroups is 1.832.

The Average Range for Method RP's six subgroups is 1.900.

The control limits for the Chart for Mean Ranges are found as follows:

$$\text{Est. SD(}\bar{R}) = \frac{\text{Est. SD(R)}}{\sqrt{M}} = \frac{1.493}{\sqrt{6}} = 0.6095 \text{ units}$$

so that:

$$\bar{\bar{R}} \pm 3\,[\,\text{Est. SD(}\bar{R})\,] = 1.974 \pm 3\,(0.6095) = 0.146 \text{ to } 3.803 \text{ units.}$$

Method	H	G	RP
Average Ranges	2.192	1.832	1.900

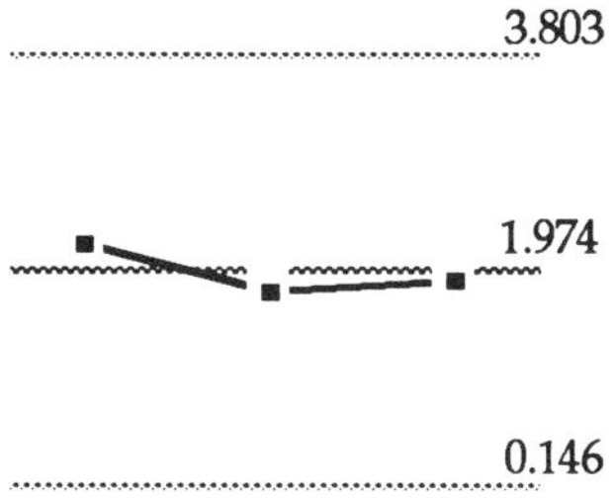

3.803

1.974

0.146

Figure 9.2: EMP for Method Inconsistency: Polymer 62S Data

Figure 9.2 shows no detectable difference in the Test-Retest Error for the three different test methods, which is in itself a surprise. The Probable Error of ± 1.17 units defines the effective resolution for all three methods. Thus, it certainly makes little sense to record the Gum Activity values to two decimal places, as was done with both Method H and Method G. One decimal place is more than sufficient for these measurements.

The Average Chart in Figure 9.1 suggests a possible Bias Effect due to the measurement method. A Main Effect Chart will check for the presence of such a Bias Effect.

EMP Study for Method Bias Effect

From the Basic EMP Study using k = 18 subgroups of size n = 2, the Grand Average is 20.934, the Average Range is 1.9744, and the estimate of the standard deviation of Test-Retest Error is 1.75 units.

The averages for each method are:
Method H averaged 22.8375 units,
Method G averaged 20.0975 units, and
Method RP averaged 19.8667 units.

Given that each of these averages is based upon 12 measurements, the estimated Test-Retest Error for these averages is:

$$\hat{\sigma}_{e\,aver} = \frac{\hat{\sigma}_e}{\sqrt{N}} = \frac{1.75}{\sqrt{12}} = 0.505 \text{ units.}$$

Therefore, the control limits for a Main Effect Chart will be:

$$\overline{\overline{X}} \pm 3\,\hat{\sigma}_{e\,aver} = 20.934 \pm 3\,(0.505) = 19.42 \text{ to } 22.45.$$

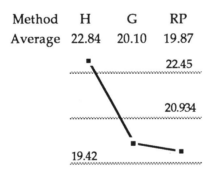

Figure 9.3: EMP for Method Bias: Polymer 62S Data

Figure 9.3 shows a detectable difference between the averages for the 3 Methods. Thus, we estimate the Method Biases to be:
Method H Bias (from the Grand Average) = + 1.9 units,
Method G Bias (from the Grand Average) = - 0.8 units,
Method RP Bias (from the Grand Average) = - 1.1 units.

Thus, with these biases one can adjust measurements made with the different methods to a common scale and make fair comparisons between samples of product. Of course, these adjustments would be appropriate only in the vicinity of 20 units.

EMP Study for Operator Bias Effect

Returning to the Average Chart in Figure 9.1, there is little or no evidence of any operator effect. However, for the purposes of illustrating how one may conduct several follow-up analyses, the checks for Operator Bias and Operator Inconsistency will be performed.

From the Basic EMP Study, using k = 18 subgroups of size n = 2, the Grand Average is 20.934, the Average Range is 1.9744, and the estimate of the standard deviation of Test-Retest Error is 1.75 units.

The averages for each operator are:

Operator A averaged 20.3925 units,

Operator B averaged 21.1117 units, and

Operator C averaged 21.2975 units.

Given that each of these averages is based upon 12 measurements, the estimated standard deviation for these averages will be the same as for the EMP Study for Method Bias, 0.505 units. This will give control limits for the Main Effect Chart of 19.42 to 22.45. These limits are the same as for the EMP Study for Method Bias because the number of Operators was the same as the number of Methods.

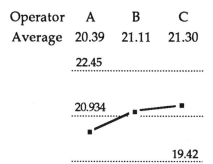

Figure 9.4: EMP for Operator Bias: Polymer 62S Data

Figure 9.4 shows no detectable difference between the Operator Averages. These data do not suggest that there is any Operator Bias in these measurements.

EMP Study for Operator Inconsistency Effects

From the Basic EMP Study using k = 18 subgroups of size n = 2, the Overall Average Range is

$$\bar{\bar{R}} = 1.9744 \text{ units}$$

which gives an estimate of the standard deviation of the Distribution of Subgroup Ranges of

$$\text{Est. SD(R)} = \frac{d_3 \bar{\bar{R}}}{d_2} = \frac{0.853 \, (1.9744)}{1.128} = 1.493 \text{ units.}$$

The average ranges for each operator are:

> The Average Range for Operator A's six subgroups is 2.055.
> The Average Range for Operator B's six subgroups is 1.843.
> The Average Range for Operator C's six subgroups is 2.025.

The control limits for the Chart for Mean Ranges are found as follows:

$$\text{Est. SD}(\bar{R}) = \frac{\text{Est. SD(R)}}{\sqrt{M}} = \frac{1.493}{\sqrt{6}} = 0.6095 \text{ units}$$

so that

$$\bar{\bar{R}} \pm 3\,[\,\text{Est. SD}(\bar{R})\,] = 1.974 \pm 3\,(0.6095) = 0.146 \text{ to } 3.803 \text{ units.}$$

Once again these limits match those found earlier because of the coincidence that the number of Operators was the same as the number of Methods.

Operator	A	B	C
Average Ranges	2.055	1.843	2.025

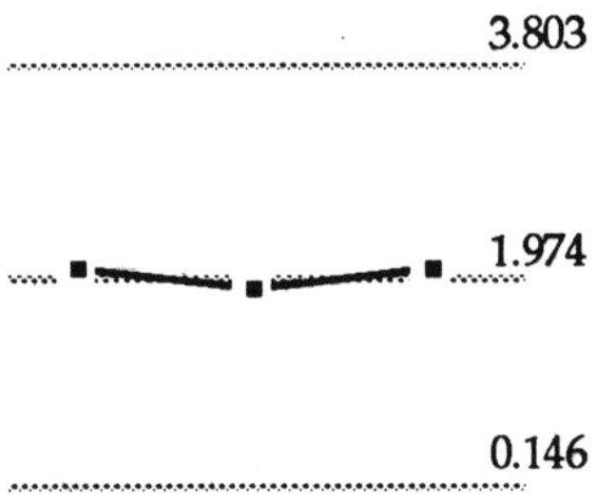

Figure 9.5: EMP for Operator Inconsistency: Polymer 62S Data

Figure 9.5 shows no evidence of any Operator Inconsistency in these data.

Summary

While the Average Chart in Figure 9.1 does show every subgroup average to be outside the control limits, this graph does not provide a measure of the relative usefulness of the measurement methods. This is due to the fact that the samples used were specially prepared samples, rather than samples drawn from the product stream. Instead of using the graph on the Average Chart, one must use the Discrimination Ratio, and this Discrimination Ratio must incorporate some measure of the product variation.

Polymer 62S was being charted using Measurement Method G. From that control chart, the estimate of σ_m^2 was found to be:

$$\hat{\sigma}_m^2 = (\bar{R}/d_2)^2 = (8.34 / 1.128)^2 = 54.7.$$

The Range Chart in Figure 9.1 provides an estimate of σ_e^2 of 3.0625.

Combining these estimates one may obtain a Discrimination Ratio of:

$$D_R = \sqrt{\frac{2\,(54.7)}{3.0625} - 1} = 5.89$$

which suggests that these Gum Activity measurements will be fairly discriminatory for Product 62S.

Thus, the three methods for measuring the Gum Activity for Polymer 62S have been found to have systematically different averages, even though they display a consistent amount of Test-Retest Error. The Bias Effects for the three methods have been estimated, which will allow for a fair comparison between measurements made with different methods. There do not appear to be any detectable Operator Effects with these measurement methods. Finally, given the Discrimination Ratio of 5.9, the three methods all appear to be satisfactory for measuring Polymer 62S.

Chapter Ten

The Compression Test Data

The EMP Study described in this chapter contains (1) data which had to be examined prior to the EMP analysis, (2) apparent Operator Effects, and (3) a measurement process that was totally inadequate to the job of detecting product variation.

A common test procedure for checking polyurethane foams is a compression test. A ram with a surface area of 50 square inches is placed on the foam sample, and then forced down into the foam a fixed distance. The force, in pounds, required to move the ram this specified distance is the measurement recorded. The greater the force, the "stiffer" the foam.

This procedure was commonly used to evaluate many different types of products. Based on these test values many manufacturers would make adjustments to their production processes. However, at one plant, some engineers decided that the test method needed to be examined. They collected samples of one foam pad (Pad 18-11) from the production line on each of five consecutive days of operation. These five samples were then tagged with an ID number to avoid confusion. Next they began to perform the compression test on each of these five parts, using the two different operators who routinely performed this test.

Since repeated compression of a polyurethane foam pad can cause the pad to take a "compression set" each of these five pads were allowed to "relax" for 24 hours between successive tests. Therefore, Round One of the data collection cycle consisted of Operator A testing the 5 pads on Day 1, and Operator B testing the 5 pads on Day 2. Round Two consisted of Operator A then retesting the 5 pads on Day 3, and Operator B retesting the 5 pads on Day 4. Finally, Round Three was performed in the same manner on Days 5 and 6. Thus, the data collected for this Basic EMP Study consisted of k = 10 subgroups of size n = 3. These data are shown in Table 10.1.

Table 10.1: Data for Compression Test EMP Study

Operator	A					B				
Pad	1	2	3	4	5	1	2	3	4	5
Round One	28.2	27.0	28.0	27.4	27.4	27.8	27.4	27.6	27.4	28.8
Round Two	27.6	27.4	27.8	27.0	28.0	27.8	28.2	29.0	28.4	30.2
Round Three	29.6	29.6	30.4	29.4	29.6	29.6	29.4	30.2	30.0	30.0

Since the possibility existed that the pads could take a "compression set," it was decided to first examine the data by Rounds. Three subgroups of size 10 were formed, with Round One being the first subgroup, Round Two being the second subgroup, and Round Three being the third subgroup. These data were then plotted on an ordinary Average and Range Chart. This chart is shown in Figure 10.1.

Round	I	II	III
Averages	27.70	28.14	29.78
Ranges	1.8	3.2	1.0

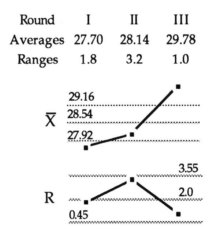

Figure 10.1: Control Chart for Compression Set

The Average Chart in Figure 10.1 gives clear evidence of a shift in the Compression Test values. The values obtained on Round Three are definitely higher that the others. This is the type of shift that should occur when the parts take a "compression set." This finding will have an impact upon how the EMP Study will be conducted.

Table 10.1 shows the original data arranged in k = 10 subgroups of size n = 3. However, with this arrangement, the variation from Round-to-Round would be included within each subgroup. Since the guideline for EMP Studies is to arrange the subgroups so that, to the greatest extent possible, only Test-Retest Error occurs within the subgroups, the arrangement in Table 10.1 should not be used. Therefore, the deletion of the data from Round Three is advisable prior to performing an EMP Study.

Following the deletion of Round Three, Rounds One and Two are checked for consistency by means of the control chart shown in Figure 10.2.

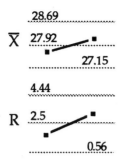

Figure 10.2: Revised Control Chart for Compression Set

Rounds One and Two do not display any obvious inconsistency on the chart in Figure 10.2. Thus, these two rounds were used for the following EMP Study.

This check for consistency from round to round prior to the EMP Study should be used whenever the experimenter suspects that there might be a time effect upon the readings. Of course, if such time effects are inherently part of the measurement process, then they might properly be made part of the EMP Study by isolating them between the subgroups. However, if such effects are due to fatigue of the parts being measured, or they are some other type of nuisance effect that has nothing to do with the purpose of the EMP Study, then they should be removed from the data prior to the EMP analysis.

EMP Study for Compression Test Data

Operator	A					B				
Pad Number	1	2	3	4	5	1	2	3	4	5
Round One	28.2	27.0	28.0	27.4	27.4	27.8	27.4	27.6	27.4	28.8
Round Two	27.6	27.4	27.8	27.0	28.0	27.8	28.2	29.0	28.4	30.2
Averages	27.9	27.2	27.9	27.2	27.7	27.8	27.8	28.3	27.9	29.5
Ranges	0.6	0.4	0.2	0.4	0.6	0.0	0.8	1.4	1.0	1.4

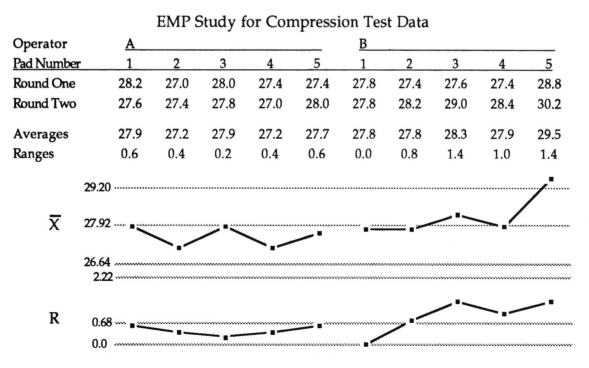

Figure 10.3: Average and Range Chart for Basic EMP Study: Compression Test Data

The EMP Control Chart in Figure 10.3 suggests several things:

(1) the compression test measurements do not appear to be very discriminating for Foam Pad 18-11;

(2) the standard deviation for Test-Retest Error is estimated to be 0.603 pounds;

(3) there may be an Operator Bias Effect; and

(4) there may be an Operator Inconsistency Effect.

The issue of the possible Operator Effects will be considered first.

EMP Study for Operator Inconsistency

From the Basic EMP Study using k = 10 subgroups of size n = 2, the Overall Average Range is

$$\bar{\bar{R}} = 0.68 \text{ pounds.}$$

which gives an estimate of the standard deviation of the Distribution of Subgroup Ranges of

$$\text{Est. SD(R)} = \frac{d_3 \bar{\bar{R}}}{d_2} = \frac{0.853 \, (0.68)}{1.128} = 0.5142 \text{ pounds.}$$

The average ranges for each operator are:

The Average Range for Operator A's five subgroups is 0.44 pounds.

The Average Range for Operator B's five subgroups is 0.92 pounds.

The control limits for the Chart for Mean Ranges are found as follows:

$$\text{Est. SD}(\bar{R}) = \frac{\text{Est. SD(R)}}{\sqrt{M}} = \frac{0.5142}{\sqrt{5}} = 0.2300 \text{ pounds}$$

from this we get:

$$\bar{\bar{R}} \pm 3 \, [\, \text{Est. SD}(\bar{R}) \,] = 0.68 \pm 3 \, (0.2300) = 0.0 \text{ to } 1.37 \text{ pounds.}$$

Operator	A	B
Average Ranges	0.44	0.92

1.37 ·····································
0.68 ·············· ━━ ■ ·········
0.0 ·····································

Figure 10.4: EMP for Operator Inconsistency: Compression Test Data

Even though the two operators appear to have considerably different average ranges, the difference is still small enough to be due to chance alone. Figure 10.4 shows no evidence of Operator Inconsistency.

EMP Study for Operator Bias

From the Basic EMP Study using k = 10 subgroups of size n = 2, the Grand Average is 27.92 pounds, the Average Range is 0.68 pounds, and the estimate of the standard deviation of Test-Retest Error is 0.603 pounds.

The two operators in the study each made 15 measurements. However, we are using only the first 10 measurements in this analysis. For these first ten measurements Operator A averaged 27.58 pounds, while Operator B averaged 28.26 pounds.

Given that each of these averages is based upon 10 measurements, the estimated Test-Retest Error for these averages is

$$\hat{\sigma}_{e\,aver} = \frac{\hat{\sigma}_e}{\sqrt{N}} = \frac{0.603}{\sqrt{10}} = 0.191 \text{ pounds.}$$

Therefore, the control limits for a Main-Effect Chart will be:

$$\overline{\overline{X}} \pm 3\,\hat{\sigma}_{e\,aver} = 27.92 \pm 3\,(0.191) = 27.35 \text{ to } 28.49 \text{ pounds.}$$

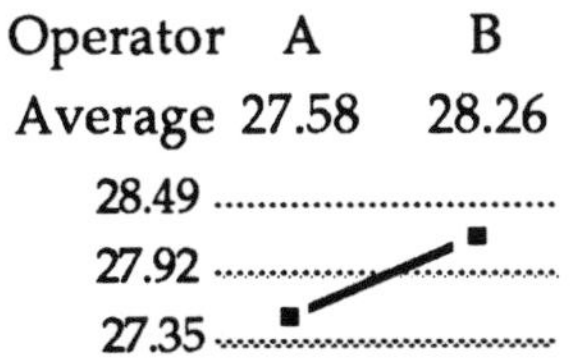

Figure 10.5: EMP for Operator Bias: Compression Test Data

Figure 10.5 shows no detectable difference between the Operator Averages. These data do not suggest that there is any Operator Bias in these measurements.

Careful examination of the data in the order in which they were obtained will suggest a reason why the chart in Figure 10.3 looked like there might be an Operator Effect. The data from Round one and Round Two are arranged by days of collection in Table 10.2.

Table 10.2: Data Arranged By Day of Collection

Operator	Day	Pad 1	Pad 2	Pad 3	Pad 4	Pad 5
A	1	28.2	27.0	28.0	27.4	27.4
B	2	27.8	27.4	27.6	27.4	28.8
A	3	27.6	27.4	27.8	27.0	28.0
B	4	27.8	28.2	29.0	28.4	30.2

Scanning the columns of Table 10.2, Four of the five pads are seen to have a notable increase in the compression value on Day 4. Thus, the apparent shift for Operator B on both the Average and Range Chart of Figure 10.3 may be due to the beginnings of the same phenomenon which led to the deletion of the data for Round Three.

Finally, since the five pads were selected over five different days of production, they should provide a reasonable estimate of the product variation. Using the averages for each pad, the Discrimination Ratio is found as follows: the standard deviation for Test-Retest Error gives:

$$\hat{\sigma}_e^2 = 0.36341.$$

The averages for the five pads used are:

$$27.85, \quad 27.50, \quad 28.10, \quad 27.55, \quad 28.60.$$

Each of these averages is based upon $N = 4$ measurements. Using these averages we find:

$$s_{aver}^2 = 0.20325$$

giving:

$$\hat{\sigma}_m^2 = s_{aver}^2 + \frac{N-1}{N}\,\hat{\sigma}_e^2 = 0.20325 + (3/4)\,0.36341 = 0.47581.$$

From these values we find:

$$D_R = \sqrt{\frac{2\,(0.4758)}{0.3634} - 1} = 1.27.$$

This Discrimination Ratio of 1.27 suggests that these measurements cannot even discriminate between high values and low values within the Natural Process Limits. About all that they appear to be capable of detecting is that the foam pads are foam pads!

Thus, while the Compression Test values have a Probable Error of ± 0.4 pounds, they do not appear to be satisfactory for measuring Foam Pad 18-11. Because the parts will take a compression set, it is not feasible to use multiple (repeated) measurements, and single measurements are not capable of detecting product variation within the Natural Process Limits. Therefore, anyone using these measurements for Foam Pad 18-11 is likely to be reacting to random variation. Every usage of these values should be carefully reviewed and the users should be made aware of just how poor these measurements are.

Appendix

Glossary

AVER(R)   The Mean of the Distribution of Subgroup Ranges, p.8

D_3 Factor for computing the Lower Control Limit for the Range Chart using $\bar{R}$, p.8

D_4 Factor for computing the Upper Control Limit for the Range Chart using $\bar{R}$, p.8

d_2 Bias Correction Factor for Ranges, p.8

d_3 Factor relating SD(R) and SD(X), p.44

D_ρ The Discrimination Ratio, p.51

D_R The Estimated Discrimination Ratio, p.51

LCL_R The Lower Control Limit for the Range Chart, p.8

$\bar{R}$ The Average Range of a set of measurements, p.15

$\bar{\bar{R}}$ The Overall Average Range for a set of subgrouped data, p.44

ρ_I The Intraclass Correlation Coefficient, p.50

r_I The Estimated Intraclass Correlation Coefficient, p.50

R_{aver} The Range of a set of subgroup averages, p.52

σ_e Standard Deviation of Test-Retest Error for single measurement, p.20

$\hat{\sigma}_e$ Estimate of Standard Deviation of Test-Retest Error for single measurement, p.14

σ_e^2 Variance of Test-Retest Error for single measurement, p.49

$\hat{\sigma}_e^2$ Estimated Variance of Test-Retest Error for single measurement, p.50

$\hat{\sigma}_{e\,aver}$ Estimated std. dev. of Test-Retest Error for average of several measurements, p.42

σ_m^2 Variance of Product Measurements, p.49

$\hat{\sigma}_m^2$ Estimated Variance of Product Measurements, p.50

σ_p^2 Variance of Product Values, p.49

$\hat{\sigma}_p^2$ Estimated Variance of Product Values, p.50

Glossary

s^2	The variance of a set of measurements, p.52
s^2_{aver}	The variance of a set of subgroup averages, p.52
$SD(R)$	The standard deviation of the Distribution of Subgroup Ranges, p.44
$SD(\bar{R})$	The standard deviation of the Average of several Subgroup Ranges, p.45
$SD(X)$	The standard deviation of the Distribution of Individual Measurements, p.8
UCL_R	The Upper Control Limit for the Range Chart, p.8
$\bar{X}$	The Average of a set of measurements, p.43
$\bar{\bar{X}}$	The Grand Average of a set of subgrouped data, p.42

Table A
Bias Correction Factors
For Estimating Standard Deviations

Subgroup Size	d_2	c_2	c_4	d_3	Subgroup Size	d_2	c_2	c_4	d_3
2	1.128	0.5642	.7979	0.8525	21	3.778	0.9638	.9876	0.7272
3	1.693	0.7236	.8862	0.8884	22	3.819	0.9655	.9882	0.7199
4	2.059	0.7979	.9213	0.8798	23	3.858	0.9670	.9887	0.7159
5	2.326	0.8407	.9400	0.8641	24	3.895	0.9684	.9892	0.7121
6	2.534	0.8686	.9515	0.8480	25	3.931	0.9695	.9896	0.7084
7	2.704	0.8882	.9594	0.8332	30	4.086	0.9748	.9915	0.6927
8	2.847	0.9027	.9650	0.8198	35	4.213	0.9784	.9927	0.6799
9	2.970	0.9139	.9693	0.8078	40	4.322	0.9811	.9936	0.6692
10	3.078	0.9227	.9727	0.7971	45	4.415	0.9832	.9943	0.6601
11	3.173	0.9300	.9754	0.7873	50	4.498	0.9849	.9949	0.6521
12	3.258	0.9359	.9776	0.7785	60	4.639	0.9874	.9957	0.6389
13	3.336	0.9410	.9794	0.7704	70	4.755	0.9892	.9963	0.6283
14	3.407	0.9453	.9810	0.7630	80	4.854	0.9906	.9968	0.6194
15	3.472	0.9490	.9823	0.7562	90	4.939	0.9916	.9972	0.6118
16	3.532	0.9523	.9835	0.7499	100	5.015	0.9925	.9975	0.6052
17	3.588	0.9551	.9845	0.7441					
18	3.640	0.9576	.9854	0.7386					
19	3.689	0.9599	.9862	0.7335					
20	3.735	0.9619	.9869	0.7287					

Unbiased estimators of various Standard Deviations will be obtained from any of the following:

$$\text{An Unbiased Est. of SD(X) is } \frac{s_n}{c_2} \text{ or } \frac{\bar{s}_n}{c_2}$$

$$\text{An Unbiased Est. of SD(X) is } \frac{s}{c_4} \text{ or } \frac{\bar{s}}{c_4}$$

$$\text{An Unbiased Est. of SD(X) is } \frac{R}{d_2} \text{ or } \frac{\bar{R}}{d_2}$$

$$\text{An Unbiased Est. of SD(R) is } \frac{d_3 \bar{R}}{d_2}$$

Table B
Control Chart Factors

For any number of subgroups of size n with

Grand Average, $\overline{\overline{X}}$,	or	Grand Average, $\overline{\overline{X}}$,
and		and
Average Range, $\overline{R}$,		Average Standard Deviation, $\overline{s}$

Subgroup Size n	A_2	D_3	D_4		Subgroup Size n	A_3	B_3	B_4
2	1.880	--	3.268		2	2.659	--	3.267
3	1.023	--	2.574		3	1.954	--	2.568
4	0.729	--	2.282		4	1.628	--	2.266
5	0.577	--	2.114		5	1.427	--	2.089
6	0.483	--	2.004		6	1.287	0.030	1.970
7	0.419	0.076	1.924		7	1.182	0.118	1.882
8	0.373	0.136	1.864		8	1.099	0.185	1.815
9	0.337	0.184	1.816		9	1.032	0.239	1.761
10	0.308	0.223	1.777		10	0.975	0.284	1.716

$$UCL_{\overline{x}} = \overline{\overline{X}} + A_2 \overline{R} \qquad\qquad UCL_{\overline{x}} = \overline{\overline{X}} + A_3 \overline{s}$$

$$CL_{\overline{x}} = \overline{\overline{X}} \qquad\qquad CL_{\overline{x}} = \overline{\overline{X}}$$

$$LCL_{\overline{x}} = \overline{\overline{X}} - A_2 \overline{R} \qquad\qquad LCL_{\overline{x}} = \overline{\overline{X}} - A_3 \overline{s}$$

$$UCL_R = D_4 \overline{R} \qquad\qquad UCL_s = B_4 \overline{s}$$

$$CL_R = \overline{R} \qquad\qquad CL_s = \overline{s}$$

$$LCL_R = D_3 \overline{R} \qquad\qquad LCL_s = B_3 \overline{s}$$

And for $n > 10$:

$$A_2 = \frac{3}{d_2 \sqrt{n}} \qquad\qquad A_3 = \frac{3}{c_4 \sqrt{n}}$$

$$D_3 = \left[1 - \frac{3 d_3}{d_2} \right] \qquad\qquad B_3 = \left[1 - \frac{3}{\sqrt{2(n-1)}} \right]$$

$$D_4 = \left[1 + \frac{3 d_3}{d_2} \right] \qquad\qquad B_4 = \left[1 + \frac{3}{\sqrt{2(n-1)}} \right]$$

EMP Studies for Bias Effects:
Worksheet for Main-Effect Chart

Step I. Information from the Basic EMP Study:

(a) Product Studied: _____   (b) Measurement Studied: _____

(c) Grand Average: _____   (d) Average Range: _____

(e) Subgroup size, n: _____   (f) Number of Subgroups, k: ____

(g) Estimate of Standard Deviation for Test-Retest Error: $\hat{\sigma}_e =$ _____

Step II. Potential Bias Effect Considered:

(a) Name of Component of Measurement Error: _____

(b) Number of Levels for this Component, L = _____

(c) Number of Observations per Level, N, (usually N = nk /L) = ____

(d) Names for Each Level of this Component:

_____ _____ _____ _____ _____ _____ _____ _____ _____ _____

(e) Averages for Each Level of this Component:

_____ _____ _____ _____ _____ _____ _____ _____ _____ _____

Step III. Compute Control Limits and Plot the Main-Effect Chart;

(a) $\hat{\sigma}_{e\,aver} = \dfrac{\hat{\sigma}_e}{\sqrt{N}} =$ _____

(b) Control Limits are $\overline{\overline{X}} \pm 3\,\hat{\sigma}_{e\,aver} =$ _____ to _____

(c) Plot the Averages in Step II.(e) against the Limits in Step III.(b).

If any of the Averages falls outside these control limits,

estimate the Bias Effect for each level of this Component of Measurement Error.

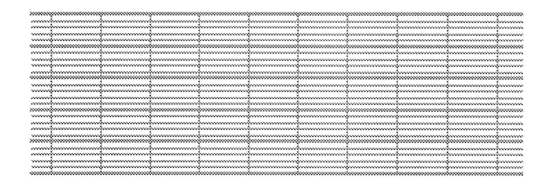

EMP Studies for Inconsistency Effects:
Worksheet for Chart for Mean Ranges

Step I. Information from the Basic EMP Study:

(a) Product Studied: _____ (b) Measurement Studied: _____

(c) Overall Average Range: $\bar{\bar{R}}$ = _____

(d) Subgroup size, n: _____ (e) Number of Subgroups, k: ____

(g) Estimate of Standard Deviation for Ranges: Est. SD(R) $= \dfrac{d_3 \bar{\bar{R}}}{d_2} =$ _____

Step II. Potential Inconsistency Effect:

(a) Name of Component of Measurement Error: _____

(b) Number of Levels for this Component, L = _____

(c) Number of Ranges for each level, M (usually M = k /L)

_____ _____ _____ _____ _____ _____ _____ _____ _____ _____ _____

(d) Names for Each Level of this Component:

_____ _____ _____ _____ _____ _____ _____ _____ _____ _____ _____

(e) Average Ranges for Each Level of this Component:

_____ _____ _____ _____ _____ _____ _____ _____ _____ _____ _____

Step III. Compute Control Limits and Plot the Chart for Mean Ranges:

(a) Est. SD($\bar{R}$) $= \dfrac{\text{Est. SD(R)}}{\sqrt{M}} =$ _____

(b) Control Limits are $\bar{\bar{R}} \pm 3\,[\,\text{Est. SD}(\bar{R})\,] =$ _____ to _____
(c) Plot the Average Ranges in Step II.(e) against the Limits in Step III.(b).
     If any of the Average Ranges falls outside these control limits, then group together
     those levels of this Component of Measurement Error which have similar Average
     Ranges, and estimate the Test-Retest Error for each grouping separately.

Worksheet for Finding Discrimination Ratios
Page One

A. Basic Information:

Product and Characteristic Measured: _____

Measurement Procedure or Gauge Used: _____

If measurement process shows inconsistency due to machine or operator effects, then note which machine and/or operator data is used in computing the Discrimination Ratio below:

_____

B. Estimate of Test-Retest Error:

The standard deviation of Test-Retest Error is obtained from either the within-subgroup variation in an EMP Study or else from a Control Chart of Repeated Measurements of the same sample part or batch.

$$\hat{\sigma}_e = \bar{R}/d_2 = \text{_____}$$

Square this to obtain:

$$\hat{\sigma}_e^2 = \text{_____}$$

C. Estimate of Product Measurement Variation:

The standard deviation of Product Measurements is estimated by the within-subgroup variation on a control chart for the product characteristic listed above. (With an XmR Chart use the measurement-to-measurement variation.)

$$\hat{\sigma}_m = \bar{R}/d_2 = \text{_____}$$

Square this to obtain:

$$\hat{\sigma}_m^2 = \text{_____}$$

(Alternate methods for estimating this variance are shown on page two of this worksheet.)

D. Estimate the Discrimination Ratio:

$$D_R = \sqrt{\frac{2\,\hat{\sigma}_m^2}{\hat{\sigma}_e^2} - 1} =$$

Worksheet for Finding Discrimination Ratios
Page Two

E. Alternate Methods of Estimating the Product Measurement Variation:

DO NOT USE either of the following methods of estimating the variation of product measurements if a regular control chart for these product measurements is available.

DO NOT USE either of the following methods if the sample parts or batches used in the EMP Study were selected for the study based on the values of the measurements (say the parts were selected so as to have some high values and some low values in the study). Such subjective selections will not provide a useful estimate of the product variation.

USE ONE OF THE FOLLOWING METHODS only if the sample parts or batches were objectively (haphazardly, systematically, or randomly) selected from the product stream without regard to the value represented by each sample part or batch.

Number of Sample Parts or Batches in the EMP Study: p = _____
Number of measurements per sample part or batch= N (where $N = nk/p$ usually) N = _____
Write the average of the N measurements for each sample part or batch below:

_____ _____ _____ _____ _____ _____ _____ _____ _____

Method 1. Compute the standard deviation, s, using these p Averages: s_{aver} = _____

Square this value to obtain s_{aver}^2 = _____
Estimate σ_m^2 by:

$$\hat{\sigma}_m^2 = s_{aver}^2 + \frac{N-1}{N}\hat{\sigma}_e^2 =$$

Method 2. Find the Range of the p Averages shown above: R_{aver} = _____
Find the d_2 value for subgroups of size p: d_2 = _____
Estimate σ_m^2 by:

$$\hat{\sigma}_m^2 = \left(\frac{R_{aver}}{d_2}\right)^2 + \frac{N-1}{N}\hat{\sigma}_e^2 =$$

Index

Modified Procedure for
Charts for Main Effects and
Charts for Mean Ranges.

While both the Charts for Main Effects and the Charts for Mean Ranges as defined herein are perfectly valid and appropriate techniques, they are both essentially set up for comparing a large number of averages. When these techniques are used with a small number of averages they can be very conservative. Therefore, the following modification is suggested as a way to remedy this deficiency.

Instead of automatically computing 3-sigma limits for the Main Effects or Mean Ranges, the user may compute H-sigma limits, using the H factors given below. The value for H will depend upon the number of averages being compared. If the Main Effect chart only has two averages, then $L = 2$ and $H = 1.82$. Likewise, if the Mean Range chart has only two average ranges, then $L = 2$ and $H = 1.82$.

L = number of averages to be compared
H = multiplier to use in computing limits.

L	2	3	4	5	6	7	8	9	10	12	15	20	24	30	40	60
H	1.82	2.38	2.61	2.75	2.87	2.94	3.01	3.07	3.12	3.20	3.28	3.39	3.45	3.53	3.62	3.73

Inspection of this table shows that the use of 3-sigma limits is very conservative when comparing two or three averages.

Using this modification, the Main Effect chart on page 83 has limits of:
$$20.934 \pm 2.38 \ (0.505) = 19.73 \text{ to } 22.14$$
so the three Operator Averages still show no detectable difference between operators.

Likewise, the Mean Range Chart on page 90 now has limits of:
$$0.68 \pm 1.82 \ (0.2300) = 0.26 \text{ to } 1.10$$
so the Average Ranges for the two operators still show no detectable difference between operators.

This modification is based upon the Analysis of Means. The multipliers given above are those for $\alpha = 0.01$ and $v = \infty$. The use of $\alpha = 0.01$ says that we are using a theoretical model to filter out 99 percent of the probable noise. Regardless of the approximate validity of the model, such conservative limits will filter out most of the probable noise, so that anything which falls outside the limits is likely to represent a detectable difference. The use of $v = \infty$ is a simplifying assumption. The effect of this assumption will be to make the multipliers H slightly smaller than they might be otherwise. In practice this effect is offset by the inherent conservatism of $\alpha = 0.01$. If one wishes to use some other level of filtering, then it would be best to use a full-blown ANOM or ANOMR approach.